First principles is a physics way of looking at the world ...
what that really means is that you boil things down to the m
fundamental truths ... and then reason up from there.

—Elon Musk

Structural Design
from First Principles

Structural Design
from First Principles

Michael Byfield

CRC Press
Taylor & Francis Group
Boca Raton London New York

CRC Press is an imprint of the
Taylor & Francis Group, an **informa** business

A SPON PRESS BOOK

CRC Press
Taylor & Francis Group
6000 Broken Sound Parkway NW, Suite 300
Boca Raton, FL 33487-2742

© 2018 by Taylor & Francis Group, LLC
CRC Press is an imprint of Taylor & Francis Group, an Informa business

No claim to original U.S. Government works

Printed by CPI on sustainably sourced paper

International Standard Book Number-13: 978-1-1385-0349-6 (Hardback)
978-1-4987-4121-7 (Paperback)

Visit the Taylor & Francis Web site at
http://www.taylorandfrancis.com

and the CRC Press Web site at
http://www.crcpress.com

Contents

Symbols and Abbreviations

a	Plate length between stiffeners or stiffened panel length
A	Cross-sectional area
A'	Area of the section above the distance y from the neutral axis
A_{chord}	Area of a chord member in a truss
A_{ct}	Area of tension zone
A_m	Area enclosed by the midline through a closed section
A'_s	Area of compression steel
A_{st}	Area of a stiffened plate including the area of the stiffeners
A_{sw}	Area of shear links
A_v	Shear area
b	Plate width
b_{eff}	Effective width
b_o	Width of the shear plane
c	Cover
C	Compression force or the equivalent uniform moment factor
C_c	Compression force in concrete
C_s	Compression force in rebar
d	Effective depth or web depth
d'	Effective depth to the compression steel
e	Eccentricity
E	Young's modulus
E_c	Young's modulus for concrete
E_s	Young's modulus for steel
f	Arch rise
f_{cd}	Design crushing stress
f_{ck}	Crushing strength at 28 days using cylinder tests
$f_{\text{ct,3days}}$	Tensile strength of the concrete 3 days after casting
f_{ctm}	Tensile strength of concrete
f_y, f_{yk}	Yield stress
G	Shear modulus
g_{k}	Dead load
h	Depth
I	Second moment of area
I_{comp}	Second moment of area of a composite beam
I_{eff}	Second moment of area of a web within the effective width
I_{st}	Second moment of area of a stiffened panel
I_t	Torsional constant
I_{truss}	Second moment of area of truss

I_w	Warping constant
I_y	Second moment of area about the strong (y–y) axis
I_z	Second moment of area about the weak (z–z) axis
k	Effective length factor or buckling coefficient
K	Wobble factor
k_{st}	Stiffened panel buckling coefficient
L	Span
L_b	Length of stiff bearing
L_{cr}	Effective length
M	Moment (applied)
$M_{b,Rd}$	Lateral torsional buckling design moment
M_{dl}	Moment due to dead load (at transfer)
M_{Ed}	Applied (ultimate limit state, ULS) moment
M_{el}	Elastic moment capacity (also $M_{el,Rd}$)
$M_{el,y}$	Elastic moment capacity about the y–y axis
$M_{el,z}$	Elastic moment capacity about the z–z axis
M_{il}	Moment due to imposed load
$M_{L/4}$	Design moment at the quarter span points for an arch
$M_{pl,Rd}$	Plastic moment capacity
M_{Rd}	Moment capacity
M_u	Maximum moment before compression steel is needed
M_y	Applied moment about the strong (y–y) axis
M_z	Applied moment about the weak (z–z) axis
N	Axial force (applied)
$N_{b,chord}$	Compression strength of chord members in truss
$N_{b,Rd}$	Design buckling force (in the absence of moments)
N_{cr}	Elastic critical buckling force
$N_{cr,y}$	Elastic critical force with buckling about the y axis
$N_{cr,z}$	Elastic critical force with buckling about the z axis
N_{crush}	Crushing strength
N_{Ed}	Applied axial load (ULS)
$N_{L/4}$	Design axial force at the quarter span points for loading of an arch
N_{local}	Local buckling force
$N_{pl,Rd}$	Crushing force
P	Point load (sometimes referred to as the *prestressing force*)
q	Shear flow
q_k	Imposed load
R	Support reaction or radius of Mohr's circle or restraint factor or resistance
s	Shear link spacing
S	Swept length or crack spacing or load
s_k	Characteristic snow load
t	Web thickness or plate thickness
T	Torsional moment or tensile force or temperature
V	Shear force
V_{Ed}	ULS shear force
$V_{pl,Rd}$	Design (plastic) shear strength
V_{Rd}	Design shear strength
w	Load or crack width
w_{dl}	Dead load
w_{eff}	Effective width

W_{el}	Elastic section modulus
$W_{el,y}$	Elastic section modulus about the y–y axis
$W_{el,z}$	Elastic section modulus about the z–z axis
w_k	Characteristic wind load
W_{pl}	Plastic section modulus
w_{sls}	Serviceability limit state load
w_{steel}	Self-weight of steel beam
w_{uls}	ULS load
x	Neutral axis depth or distance from a jack to where a tendon force is required
z	Lever arm distance
Z_B	Elastic section modulus required to give the bending stress at the bottom fibres of the section
z_c	Lever arm for the concrete force
z_{s1}, z_{s2}	Levers arms for the rebar
Z_T	Elastic section modulus required to give the bending stress at the top fibres of the section

Greek symbols

α	Amplification of moments factor or short-term loss factor or coefficient of linear expansion
α_y	Moment amplification factor for the y–y axis
α_z	Moment amplification factor for the z–z axis
β	Total loss factor
γ_c	Partial safety factor for concrete
γ_F	Partial safety factor for load
γ_M	Partial safety factor for material properties
γ_s	Partial safety factor for yield stress
Δ, δ	Deflection
δ_1	Uplift due to the total drape of the tendon
δ_2	Downward movement due to support moments
δ_{dl}	Deflection due to dead load
δ_{il}	Deflection due to imposed load
δ_{total}	Total deflection
ε	Strain
ε_c	Concrete strain
ε_{free}	Free shrinkage strain
ε_{pre}	Pre-tensioning strain
ε_r	Restrained shrinkage strain
$\varepsilon_{shrinkage}$	Shrinkage strain
$\varepsilon_{specific}$	Specific creep strain
ε_x	Strain in an element according to Hooke's law
ε_y	Yield strain
θ	Angle of twist or angle inclination or slope
λ	Slenderness
μ	Coefficient of friction
ν	Poisson's ratio or strength reduction factor
ρ_p	Reinforcement ratio
$\rho_{p,eff}$	Effective reinforcement ratio
σ	Compression or tension stress
σ_B	Stress in the bottom of the beam

σ_c	Average stress in the concrete along the line of the tendons
σ_{cr}	Elastic critical buckling stress
σ_{il}	Tensile stress due to imposed loading
σ_{loss}	Loss of stress in the steel tendons
σ_{max}	Maximum stress
$\sigma_{max,sls}$	Maximum permissible compression stress at the SLS
$\sigma_{max,t}$	Maximum permissible compression stress at transfer
$\sigma_{min,sls}$	Minimum permissible (tension) stress at the SLS
$\sigma_{min,t}$	Minimum permissible (tensile) stress at transfer
σ_{Rd}	Design value for the buckling stress or design stress for a plate
σ_T	Stress in the top of the beam
σ_t	Principle tensile stress
σ_x	Stress in the x direction
σ_y	Yield stress or stress in the y direction
τ	Shear stress
τ_{cr}	Elastic critical buckling shear stress
τ_{Ed}	Applied shear stress
τ_{Rd}	Design value of shear stress
τ_y	Yield shear stress
ϕ	Internal angle of web members or rebar diameter

Abbreviations

CCC	Compression compression compression (node)
CCT	Compression compression tension (node)
CTT	Compression tension tension (node)
FEA	Finite element analysis
FoS	Factor of safety
LTB	Lateral torsional buckling
PSC	Prestressed concrete
RC	Reinforced concrete
SLS	Serviceability limit state
STM	Strut and tie model
UDL	Uniformly distributed load
ULS	Ultimate limit state

Foreword

I've been a structural engineer now for the best part of 50 years. Over that time, our profession has moved through a revolution at a bewildering pace. When I started out, my sole calculation equipment was a pencil, a rubber and a slide rule. All drawings were prepared by hand in ink. Skills in neat lettering were essential. I've worked through the era of the first pocket calculator, the first programmable calculator, simple desktops and into the era of super computers offering incredible facilities to predict structural performance. During that same 50 years, the amount of published work on structural behaviour has been phenomenal. So as a profession, we can now rely on a huge knowledge base.

These developments are a huge boon. However, they bring with them enormous challenges. Students today have to learn most of what I had to learn 50 years ago plus things discovered since. So in many ways, it's harder for today's generation than it was for me. Over my career, I've had the good fortune to work on just about every form of structure imaginable, from simple lintels to nuclear power plants. I've worked on high-rise, all kinds of roller coasters, the London Eye and Wembley Stadium. I've had to design for a whole range of conditions, from the simplest static loading through to predicting complex dynamic performance and taken on the responsibilities of preventing building collapse in earthquakes. However, I was able to work up those skills gradually over a long period.

In facing up to those challenges faster, today's generation have computers to help. However, with that advantage comes the concern of having confidence in the validity of computer predictions. Paradoxically, the programs that help us most (for complex structures) raise the most concern, since the designers involved can lose all 'feel' for what the answer should be. Chapter 9 of this book highlights the sobering Sleipner disaster as an example: I could tell tales of many more.

Hence, my advice to all aspiring structural engineers is to make sure your training includes developing a thorough understanding of the basics of how structures perform under stress, before you get lost in equations. Make sure you have the skills to check even complex structures by hand, so you can independently verify that complex strength predictions are of the right order. It is not necessary to be precise. Indeed, any presumption that computer output is 'accurate' is itself a fiction. If you read this book, absorb its timeless principles and work your way through the examples, you will learn a great deal and it will serve you well in your career.

Allan P. Mann, BSc, PhD, FIStructE, FREng

Preface

When I began my career in academia, I taught students how to use the British Standards to design members. The Eurocodes were introduced and I was faced with the challenge of teaching some very complicated design methods to students who had only a basic understanding of mechanics. I decided I needed to teach students the first principles and began out of necessity to write lecture notes that turned into this book. During this process, I found it was possible to tackle some problems that would ordinarily be outside the scope of traditional undergraduate courses. Topics like the design of long-span bridges, which if treated from first principles, become quite easy to understand.

I have not ignored the codes. In fact, I have used the Eurocode safety factors and notation throughout. However, the formulae are in many cases quite different. My intention has been to convey a firm understanding—not of the current codes themselves—but of the underpinning principles. I hope this will help young engineers to face the future, whichever design codes they use.

I would like to express my thanks to Louise (my wife) and also John (my father), who patiently proof read the manuscript. I would also like to thank Allan Mann, who found the time to read the entire manuscript and write the Foreword to this book. I would also like to thank my Editor, Tony Moore, for his help throughout.

Author

Dr. Michael Byfield, BEng, PhD, MIStructE, MICE, CEng, is a lecturer in structural engineering at Southampton University (UK) and runs a small structural engineering consultancy.

Chapter 1

Limit state design

This chapter outlines the philosophy behind what Europeans call *limit state design* and Americans call *load and resistance* (safety) *factor design*. This is the method used by regulators to write modern structural design codes and it involves the application of partial safety factors to load and resistance. The 'limit state' is a condition beyond which a structure no longer fulfils the design intent, and there are two types:

1. *Serviceability limit state (SLS) design*: The structure must be fit for purpose under working loads. For most situations, this means the structure must *remain elastic* and not *deflect excessively* when supporting *unfactored loads*.
2. *Ultimate limit state (ULS) design*: The structure must be strong enough to support loads increased using (partial) safety factors. Unlike SLS design, the engineer can utilise the full plastic design strength *if* a material is ductile. For example, plastic design is allowed for some steel members, whereas brittle materials, such as wood, must be designed using elastic principles.

1.1 PARTIAL SAFETY FACTORS

The objective of ULS design is to ensure that

Design strength ≥ Design load

which is expressed as

$$\frac{R}{\gamma_M} \geq S \times \gamma_F \tag{1.1}$$

where
  R is the resistance (strength).
  γ_M is the material partial safety factor.
  S is the estimate for load.
  γ_F is the load partial safety factor.

The partial safety factors are normally based on proven work over many years. When new design equations are added to codes, they are tested for accuracy in the laboratory. It is important to remember that design resistances and loads are only approximations. The statistical uncertainty is modelled using the log-normal probability distribution function for both load and resistance, a process that is illustrated in Figure 1.1. The partial safety factors are selected to ensure that probability of failure is very small.

Lists of partial safety factors (γ-factors) are shown in Tables 1.1 and 1.2. It can be seen that different factors are applied to different types of load and materials. Equation 1.1 is therefore a simplification, because more than one factor is applied to load and, in the case of reinforced concrete, material factors are applied to both steel and concrete separately. In the case of structural steelwork, the material factor can be neglected from resistance calculations because it is set at 1.0.

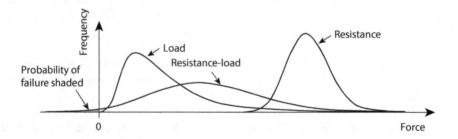

Figure 1.1 Probability distributions of resistance, load and resistance-load.

Table 1.1 Eurocode partial safety factors for loads

Load	Partial safety factor, γ_F
Dead, g_k	1.35 or 1.0
Imposed, q_k	1.5
Wind, w_k	1.5
g_k, q_k and w_k combined	1.35

Table 1.2 Eurocode partial safety factors for materials

Material strength	Partial safety factor, γ_M
Structural steelwork	1.0
Steel rebar	1.15
Concrete	1.5
Timber	1.3

1.2 CALCULATION OF LOADS

The two types of loading are gravity and non-gravity. Gravity loads include

Dead load, g_k (e.g. structural self-weight)
Imposed load, q_k (everything that can be removed, e.g. furniture or vehicles)
Snow, s_k

and non-gravity loads include

Wind, w_k
Seismic forces
Accidental loads, such as impacts

The SLS load (w_{sls}) is simply the combination of these loads; for example, the most common SLS load combination is

$$w_{sls} = g_k + q_k \tag{1.2}$$

The ULS load (w_{uls}) is calculated using a combination of the loads and the relevant partial safety factors from Table 1.1. Some common load combinations are

$$w_{uls} = 1.35g_k + 1.5q_k \tag{1.3}$$

$$w_{uls} = 1.35 \times (g_k + q_k + w_k) \tag{1.4}$$

$$w_{uls} = 1.0g_k + 1.5w_k \tag{1.5}$$

Example 1.1: Simple steel beam

A steel beam weighs 74 kg/m, spans 12 m and supports an imposed load of 5 kN/m. Young's modulus is 210,000 N/mm² and the second moment of area of the beam is 32,670 cm⁴.

1. Determine the deflection under SLS loading.
2. Determine the ULS shear force and design moment.

1. The dead load of the steel beam is

$$w_{steel} = m \times g = 74 \times 9.81 \times 10^{-3} = 0.7 \text{ kN/m}$$

From Equation 1.2, the SLS load is

$$w_{sls} = g_k + q_k$$

$$w_{sls} = 0.7 + 5 = 5.7 \text{ kN/m} = 5.7 \text{ N/mm}$$

The second moment of area of the steel I-beam is provided in cm^4 and therefore needs converting:

$$I = 32670 \text{ cm}^4 = 32670 \times 10^4 \text{ mm}^4$$

The mid-span deflection in a beam subjected to a uniformly distributed load (UDL) is

$$\delta = \frac{5wL^4}{384EI} \tag{1.6}$$

Therefore,

$$\delta = \frac{5 \times 5.7 \times 12000^4}{384 \times 210000 \times 32670 \times 10^4} = 22 \text{ mm}$$

The usual deflection limit for a beam is span/360, which in this case is 33 mm; therefore, this beam has passed.

2. From Equation 1.3, the total ULS load is

$$w_{uls} = 1.35g_k + 1.5q_k$$

$$w_{uls} = 1.35 \times 0.7 + 1.5 \times 5 = 8.4 \text{ kN/m}$$

and the ULS shear force (V_{Ed}) and moment (M_{Ed}) are

$$V_{Ed} = \frac{wL}{2} \tag{1.7}$$

$$V_{Ed} = \frac{8.4 \times 12}{2} = 50.4 \text{ kN}$$

and

$$M_{Ed} = \frac{wL^2}{8} \tag{1.8}$$

$$M_{Ed} = \frac{8.4 \times 12^2}{8} = 151.2 \text{ kN.m}$$

The bending moment and shear force capacities now need checking to make sure they are higher than V_{Ed} and M_{Ed}, although these checks are left until later chapters.

Example 1.2: Slab supported by beams

Figure 1.2 shows an RC slab supported by I-section beams spanning 7 m between simple supports. The 200 mm thick slab is loaded by an imposed load of 5 kN/m², the density of reinforced concrete (RC) is 25 kN/m³ and the I-beams weigh 70 kg/m.

1. Determine the SLS load for the central beams.
2. If the I-beams have a second moment of area of 19,500 cm⁴ and Young's modulus of 210,000 N/mm², determine the mid-span deflection under SLS loading.
3. Determine the ULS design moment and shear force.

1. The dead load of a beam is

$$w_{steel} = m \times g = 70 \times 9.81 \times 10^{-3} = 0.7 \text{ kN/m}$$

The central beams each support a 3 m wide section of 0.2 m thick slab; therefore, the slab dead load is

$$w_{slab} = \text{area} \times \text{density} = 0.2 \times 3.0 \times 25 = 15 \text{ kN/m}$$

The combined dead load is

$$g_k = 0.7 + 15 = 15.7 \text{ kN/m}$$

and the imposed load is

$$q_k = 3.0 \times 5 = 15.0 \text{ kN/m}$$

From Equation 1.2, the SLS load is

$$w_{sls} = 15.7 + 15.0 = 30.7 \text{ kN/m}$$

2. From Equation 1.6, the mid-span deflection is

$$\delta = \frac{5 \times 30.7 \times 7000^4}{384 \times 210000 \times 19500 \times 10^4} = 23 \text{ mm}$$

3. From Equation 1.3, the ULS load is

$$w_{uls} = 1.35 \times 15.7 + 1.5 \times 15.0 = 43.7 \text{ kN/m}$$

Each beam assumed to support a 3 m section of slab

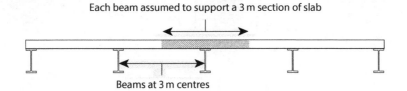

Beams at 3 m centres

Figure 1.2 Cross section through a concrete slab supported by steel beams.

From Equation 1.7, the shear force at the support is

$$V_{Ed} = \frac{43.7 \times 7}{2} = 153.0 \text{ kN}$$

and from Equation 1.8, the mid-span bending moment is

$$M_{Ed} = \frac{43.7 \times 7^2}{8} = 267.7 \text{ kN.m}$$

Example 1.3: Wind loading to a tall building

The 69 m wide, 350 m high tower shown in Figure 1.3 is subjected to a wind load of 1.4 kN/m².

1. If sideways sway is prevented only by the concrete core shown in Figure 1.3b, determine the base moment and shear force developed by the ULS wind load.
2. Determine the maximum deflection under SLS wind loading if Young's modulus for the concrete core is 35,000 N/mm².
3. Determine the deflection if the columns and the core are connected together by outrigger trusses, ensuring that they act compositely to resist sideways sway forces (see Figure 1.3c).
4. If the maximum allowable wind-induced sway is height/500, determine if the building satisfies the SLS condition for sideways movement.

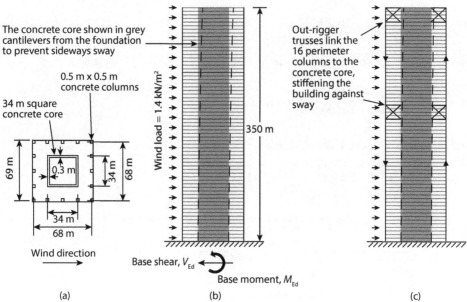

Figure 1.3 Tall building design.

1. The wind load partial safety factor is 1.5 (from Table 1.1) and the building is 69 m wide; therefore, the building develops the following load per m of height:

$$w_{uls} = \gamma_F \times w_k = 1.5 \times 1.4 \times 69 = 145 \text{ kN/m}$$

The central core works as a cantilever extending from the substructure to resist wind loading and the base shear is

$$V_{Ed} = w \times L = 145 \times 350 = 50.75 \times 10^3 \text{ kN}$$

and the base moment is

$$M_{Ed} = \frac{wL^2}{2}$$

$$M_{Ed} = \frac{145 \times 350^2}{2} = 8.88 \times 10^6 \text{ kN.m}$$

2. The second moment of area of the central concrete core is calculated in the same way as that of a hollow rectangular cross section, i.e.,

$$I_{core} = \frac{34 \times 34^3}{12} - \frac{(34 - 0.6)^4}{12} = 7655 \text{ m}^4$$

and the SLS wind load per m height of the tower is

$$w_{sls} = 1.4 \times 69 = 96.6 \text{ kN/m}$$

During the deflection calculation, all the units need to be consistent; therefore, Young's modulus is converted:

$$E = 35000 \text{ N/mm}^2 = 35000 \times 10^3 \text{ kN/m}^2$$

The maximum deflection is calculated using the equation for a cantilever supporting a uniformly distributed load, i.e.,

$$\delta = \frac{wL^4}{8EI}$$

$$\delta = \frac{96.6 \times 350^4}{8 \times 35000 \times 10^3 \times 7655} = 0.7 \text{ m} \tag{1.9}$$

3. The outer columns are fixed in position using the cross bracing shown in Figure 1.3c. This will stiffen the building, because the outer columns and core will work together to resist sideways sway. The effective second moment of area of the building ($I_{building}$) can be estimated using the parallel axis theorem:

$$I_{building} = I_{core} + \sum \left(I_{column} + \text{Area}_{column} \cdot r^2 \right) \tag{1.10}$$

The second moment of area of each column is

$$I_{\text{column}} = \frac{0.5 \times 0.5^3}{12} = 0.0052 \text{ m}^4$$

Using Equation 1.10

$$I_{\text{building}} = 7655 + 16 \times 0.0052 + \left(10 \times 0.5^2 \times 34^2 + 4 \times 0.5^2 \times 17^2 + 2 \times 0.5^2 \times 0^2\right) = 10834 \text{ m}^4$$

and from Equation 1.9 the deflection is

$$\delta = \frac{96.6 \times 350^4}{8 \times 35 \times 10^6 \times 10834} = 0.5 \text{ m}$$

It should be noted that this approach will underestimate deflections, because it does not account for the stretching and squashing of the members due to tension and compression, although the result is good enough for checking the output from a computer-based solution or for providing a guide to a likely response during the early phase of a design process.

4. The maximum allowable deflection is

$$\delta = \frac{L}{500} = \frac{350}{500} = 0.700 \text{ m}$$

Therefore, the tower is approximately satisfactory in terms of deflection, both with and without the outrigger truss, which has reduced the deflection from 0.7 to 0.5 m.

1.3 FACTOR OF SAFETY

The factor of safety (FoS) is not a formal part of limit state design process and it is completely different from the partial safety factor. The FoS provides the engineer with a measure of how much stronger a structure is than is required to support the basic working loads, i.e.,

$$\text{Factor of safety} = \frac{\text{Design strength}}{\text{Working load}} \tag{1.11}$$

This is useful because it quantifies the amount of overload possible before failure. The FoS can be used to identify the first mode of failure for a given structure. For example, if the FoS is calculated for every failure mode in a bridge, then the failure mode with the lowest FoS is likely to fail first. If that mode was considered to be sudden in nature, then the designer may decide to strengthen that part of the structure in order to ensure that a ductile failure mode would become critical. This is because ductile failures are less dangerous than brittle or buckling failure modes. The working load is the load calculated in the absence of partial safety factors; therefore, it is the same as the serviceability limit state load.

1.4 PATTERN LOADING

Figure 1.4a shows the combination of dead + imposed loads for a simply supported, single span beam. This combination provides the maximum moments and shears and further combinations

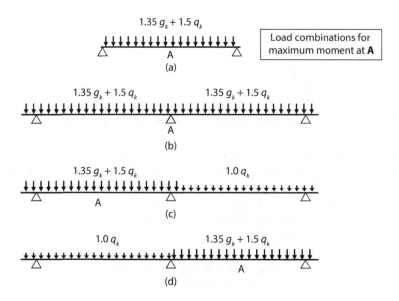

Figure 1.4 Load combinations for beams resisting dead (g_k) and imposed (q_k) loads.

are unnecessary. However, the continuous beam shown above is more complicated, as no single combination of loads will provide the maximum moment or shear at every position. The combination of full factored loads across the entire two spans (b) provides the maximum moment at the internal support (labelled A). Unfactoring the dead load and removing the imposed load on the right-hand span provides the maximum midspan moment on the left-hand span (Figure 1.4c) and likewise for Figure 1.4d. This process is called *pattern loading*.

Wind loads need to be combined with up to two separate gravity loads (dead and imposed). In addition, wind loads can act upwards, as well as downwards or sideways. Since wind loads are often upwards due to suction, the critical load combination for wind is often unfactored dead, no imposed + fully factored wind.

Problems

Solutions to these problems are provided at https://www.crcpress.com/9781498741217

P.1.1. A beam is 200 mm wide, 300 mm deep and spans 6 m. It carries an unfactored uniformly distributed imposed load of 3 kN/m run. Reinforced concrete weighs approximately 25 kN/m³ and Young's modulus for the concrete is 20,000 N/mm².

Calculate
a. The unfactored weight of the beam per m run
b. The ULS load per m run
c. The ULS bending moment
d. The ULS shear force
Ans. (a) 1.5 kN/m, (b) 6.53 kN/m, (c) 29.4 kN.m and (d) 19.6 kN.

P.1.2. An 8 m long, simply supported beam supports a dead load of 1 kN/m and an imposed load of 2 kN/m. Determine the ULS load, bending moment and shear force. If the second moment of area of the beam is 3500 cm⁴ and Young's modulus is 210,000 N/mm², determine the maximum deflection under SLS loads.
Ans. 4.35 kN/m, 34.8 kN.m, 17.4 kN, 21.8 mm.

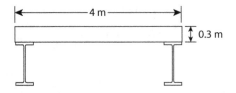

Figure 1.5 Cross section through a slab supported by two beams.

P.1.3. The 4 m wide RC slab sketched in Figure 1.5 is supported by two beams that span 8 m between simple supports. The slab is loaded by an imposed load of 5 kN/m². Each I-beam has a second moment of area of 45,000 cm⁴, Young's modulus is 210,000 N/mm² and each beam has a self-weight of 1.2 kN/m. The density of concrete is 25 kN/m³. For one of the girders only determine

a. The unfactored UDL
b. The maximum deflection under SLS dead and imposed loading
c. The ULS UDL, bending moment and shear force

Ans. (a) 26.2 kN/m, (b) 14.8 mm and (c) 36.9 kN/m, 295.2 kN.m, 147.6 kN.

Chapter 2

Steel members in flexure

Hot-rolled steel sections are rolled from hot slabs of steel, and the web and flanges are thick and stocky. They are therefore not generally prone to 'local buckling' during bending. In contrast, thin-walled sections, such as plate girders, are composed of thin plates. The compression stresses induced by bending can cause local buckling to initiate failure, and this complicates design considerably. This chapter concentrates on the calculation of the shear strength and bending strength of hot-rolled sections, although thin-walled sections are also briefly considered.

Technicians at steel mills test all rolled sections thoroughly, and any members that fail to achieve the design stress are reclassified. As a result, the actual yield stress of steel members tends to be higher than assumed during design. It is for this reason that the partial safety factor applied to steel member design is set at 1.0. Since it is equal to unity, it is not included in the equations presented in this chapter.

2.1 SHEAR STRENGTH

Von Mises yield criteria is the basis of the shear strength calculations for hot-rolled steel beams. It is popular because it provides the stress required to cause yielding in members subjected to combined bending and shear. Von Mises showed that steel will *not yield* if

$$\sqrt{\sigma_x^2 + 3\tau_{xz}^2} \leq f_y \tag{2.1}$$

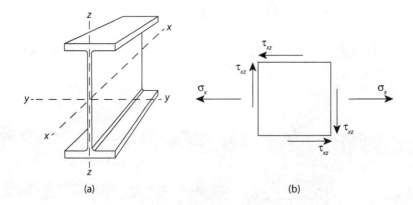

Figure 2.1 Bending and shear stresses for use with the von Mises formula. (a) Axes and (b) bending and shear stress.

where σ_x is the bending stress, τ_{xz} is the shear stress and f_y is the yield stress (see Figure 2.1). If shear is the only applied loading (i.e. $\sigma_x = 0$), then this equation shows that the shear stress required to cause yielding is

$$\tau_y = \frac{f_y}{\sqrt{3}} \tag{2.2}$$

2.1.1 Hot-rolled sections

This type of section has thick, stocky webs and these are not liable to buckling due to shear stresses. It is relatively easy to determine the shear force required to cause yielding (see Figure 2.2b). However, final failure due to shear will occur only after all the web material has fully yielded (see Figure 2.2c). In this situation, the design shear strength is approximately given by the yield shear stress (Equation 2.2) multiplied by the web area. In codes of practice, this is usually shown as

$$V_{\text{pl, Rd}} = \frac{A_v f_y}{\sqrt{3}} \tag{2.3}$$

where
  $V_{\text{pl,Rd}}$ is the design shear strength.
  A_v is the shear area.

For hot-rolled I- and H-sections (see Figure 2.3a), the shear area is

$$A_v = t \times D \tag{2.4}$$

and for square or rectangular hollow sections (see Figure 2.3b)

$$A_v = 2t \times D \tag{2.5}$$

2.1.2 Thin-walled sections

Shear stresses can cause buckling of thin plates. This is called *shear buckling* and is illustrated in Figure 2.4.

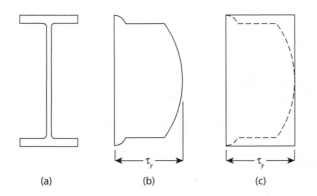

(a) (b) (c)

Figure 2.2 Shear stress distributions in hot-rolled I-sections. (a) Hot-rolled I-section, (b) elastic shear stress distribution at first yield and (c) plastic shear stress distribution assumed in the web at failure.

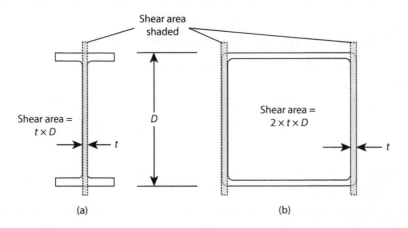

Figure 2.3 Shear area. (a) Hot-rolled I-sections and (b) hot-rolled hollow sections.

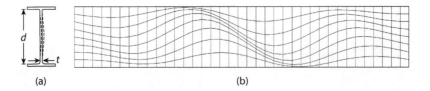

(a) (b)

Figure 2.4 Illustration of shear buckling of a thin-webbed member. (a) Shear buckling and (b) side view of web with contours showing shear buckling in a web.

This is not a problem for hot-rolled sections, but it is important for plate girders and other thin-walled sections, which can buckle at stresses well below the yield value of shear stress (Equation 2.2). The design of this type of member is described in more detail in Chapter 5, but briefly buckling will occur when the elastic critical shear stress is exceeded, i.e.,

$$\tau_{cr} = 5.34 \frac{\pi^2 E}{12(1-\upsilon^2)}\left(\frac{t}{d}\right)^2 \tag{2.6}$$

where
    τ_{cr} is the elastic critical shear stress.
    ν is the Poisson's ratio.
    E is the Young's modulus.
    t is the web thickness (Figure 2.4a).
    d is the web depth.

For steel, $\nu = 0.3$, $E = 210,000$ N/mm² and this becomes

$$\tau_{cr} = 5.34 \times 190000 \times \left(\frac{t}{d}\right)^2 \tag{2.7}$$

In simple terms, if $\tau_{cr} < \tau_y$, then failure will be by shear buckling and design should be in accordance with the mechanics of thin-walled sections described in Chapter 5. If however $\tau_{cr} > \tau_y$, then failure will be by yielding and the section can be designed using Equation 2.3.

Example 2.1: Shear strength of a hot-rolled cross section

Determine the shear strength of the hot-rolled section shown in Figure 2.5 if the yield stress is 355 N/mm².

The shear area from Equation 2.4 is

$$A_v = t \times D$$

$$A_v = 4.8 \times 177.8 = 853 \text{ mm}^2$$

Shear buckling is not a problem for hot-rolled sections; therefore, the shear strength from Equation 2.3 is

$$V_{pl,\,Rd} = \frac{A_v f_y}{\sqrt{3}}$$

$$V_{pl,\,Rd} = \frac{853 \times 355}{\sqrt{3}} \times 10^{-3} = 174 \text{ kN}$$

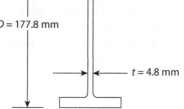

$D = 177.8$ mm

$t = 4.8$ mm

Figure 2.5 Shear area.

Example 2.2: Thin-webbed cross section

An I-section bridge girder is built up from welded plates. It has a web 1200 mm deep and 15 mm thick. If the yield stress is 355 N/mm², determine if the girder web will fail by shear buckling or yielding.

From Equation 2.2, the yield shear stress is

$$\tau_y = \frac{f_y}{\sqrt{3}}$$

$$\tau_y = \frac{355}{\sqrt{3}} = 205 \text{ N/mm}^2$$

The elastic critical shear stress from Equation 2.7 is

$$\tau_{cr} = 5.34 \times 190000 \times \left(\frac{t}{d}\right)^2$$

$$\tau_{cr} = 5.34 \times 190000 \times \left(\frac{15}{1200}\right)^2 = 158 \text{ N/mm}^2 \tag{2.8}$$

Since $\tau_{cr} < \tau_y$, this section will begin to fail by shear buckling when the shear stress reaches approximately 158 N/mm².

2.2 BENDING STRENGTH OF LATERALLY RESTRAINED BEAMS

When a beam bends, one half is thrown into compression and this can cause 'local buckling' of the flanges and web at stresses well below the yield stress. The susceptibility to local buckling is measured by the 'section classification', whereby cross sections are classified into one of four classes (see Figure 2.6). The classification is based on the web and flange width to thickness ratios (b/T and d/t). These are modified by the parameter ε. This provides tighter limits for high strength steels, since these operate at higher stresses and are therefore more prone to buckling.

Figure 2.7 shows the moment versus rotation behaviour for differing section classifications. In Class 1 and Class 2 sections, local buckling does not adversely affect strength. At the other extreme, Class 4 sections fail at stresses below the yield stress due to local buckling. The first step in determining moment capacity is to classify the cross section using the limits shown in Figure 2.6.

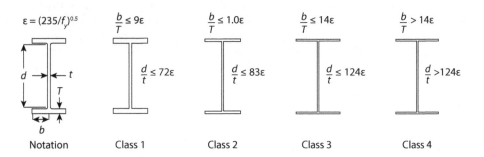

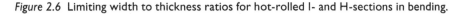

Figure 2.6 Limiting width to thickness ratios for hot-rolled I- and H-sections in bending.

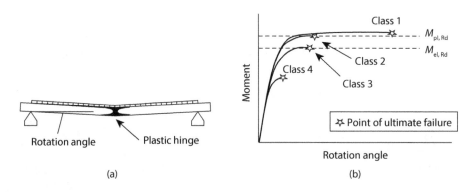

Figure 2.7 Moment versus end rotation through to failure for different section classifications. (a) Loading and (b) moment rotation behaviour.

From inspection of Figure 2.6, it can be seen that the limits for flanges are far more oner-ous than for webs. Flanges of I-beams are more prone to local buckling, because the outside edge is unrestrained against buckling. The flanges of rectangular hollow sections can toler-ate a higher b/t ratio (roughly twice) since they are restrained at each edge.

The design strengths for each class are:

Classes 1 and 2	$M_{pl,\,Rd} = f_y W_{pl}$	(2.9)
Class 3	$M_{el,\,Rd} = f_y W_{el}$	(2.10)
Class 4	The member needs to be designed as a *thin walled section* in accordance with Chapter 5	

where
$M_{pl,\,Rd}$ is the plastic moment capacity.
$M_{el,\,Rd}$ is the elastic moment capacity.
W_{pl} is the plastic section modulus.
W_{el} is the elastic section modulus.

Example 2.3: Calculate the moment capacity of a cross section

A hot-rolled I-section beam has a yield stress of 355 N/mm², a plastic section modulus of 566 cm³ (about the major axis) and is classified as a Class 1 cross section. Determine the design moment capacity.

Since this is a Class 1 cross section, Equation 2.9 defines strength. The moment capacity must be calculated in consistent units and it is easiest to work in the units of N and mm, thus

$$W_{pl,\,y} = 566 \text{ cm}^3 = 566 \times 10^3 \text{ mm}^2$$

and from Equation 2.9

$$M_{pl,\,Rd} = f_y W_{pl}$$

$$M_{pl,\,Rd} = 355 \times 566000 \times 10^{-6} = 201 \text{ kN.m}$$

The 10^{-6} in the above equation converts the solution from N.mm to kN.m.

Example 2.4: Basic beam design

A simply supported beam shown in Figure 2.8 spans 6 m between simple supports and supports a point load of 10 kN applied at midspan. Yield stress is 355 N/mm², Young's modulus is 210,000 N/mm² and the density of steel is 7700 kg/m³.

1. Determine the self-weight per m length.
2. Determine the maximum ULS bending moment and shear force.
3. Determine if the elastic moment capacity of the section is sufficient to resist the applied ULS loading.
4. Determine if the shear capacity of the section is sufficient to resist the applied ULS loading.
5. Determine the deflection under the serviceability limit state dead and imposed loads. If the maximum allowable deflection is span/200, is this beam satisfactory?

1. The cross-sectional area of the beam is

$$\text{Area} = 100 \times 150 - (100 - 2 \times 8) \times (150 - 2 \times 8) = 3744 \text{ mm}^2$$

and the beam self-weight is

$$\text{Weight} = 9.81 \times 3744 \times 10^{-6} \times 7700 \times 10^{-3} = 0.28 \text{ kN/m}$$

2. The midspan moment (M) in a beam with length (L) supporting a uniformly distributed load (w) and a centrally applied point load (P) is

$$M = \frac{wL^2}{8} + \frac{PL}{4} \tag{2.11}$$

The dead load factor is 1.35 and the imposed load factor is 1.5, therefore, the (factored) ULS moment is

$$M_{Ed} = \frac{1.35 \times 0.28 \times 6^2}{8} + \frac{1.5 \times 10 \times 6}{4} = 24.2 \text{ kN.m}$$

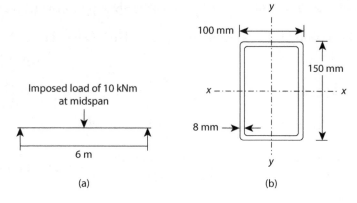

(a)

(b)

Figure 2.8 Steel beam. (a) Loading arrangement and (b) section through beam.

and the maximum shear force (V) is

$$V = \frac{wL}{2} + \frac{P}{2} \qquad (2.12)$$

Therefore, the applied shear force is

$$V_{Ed} = \frac{1.35 \times 0.28 \times 6}{2} + \frac{1.5 \times 10}{2} = 8.6 \text{ kN}$$

3. The second moment of area (I) is

$$I = \frac{100 \times 150^3}{12} - \frac{84 \times 134^3}{12} = 11.28 \times 10^6 \text{ mm}^4$$

The elastic moment capacity is calculated using the engineer's beam equation, i.e.,

$$M_{el, Rd} = \frac{f_y I}{d/2} \qquad (2.13)$$

Thus

$$M_{el, Rd} = \frac{355 \times 11.28 \times 10^6}{75} \times 10^{-6} = 53.4 \text{ kN.m}$$

Since $M_{el, Rd}$ (53.4 kN.m) > M_{Ed} (24.2 kN.m), the beam will easily support the ULS moment.

4. The shear area from Equation 2.5 is

$$A_v = 2 \times 8 \times 150 = 2400 \text{ mm}^2$$

and the shear strength from Equation 2.3 is

$$V_{pl, Rd} = \frac{2400 \times 355}{\sqrt{3}} \times 10^{-3} = 492 \text{ kN}$$

Since $V_{pl, Rd}$ (492kN) >> V_{Ed} (8.6kN), the beam will easily support the ULS shear force.

5. The midspan deflection (Δ) for a beam supporting a UDL (w) and point load (P) is

$$\Delta = \frac{5wL^4}{384EI} + \frac{PL^3}{48EI} \qquad (2.14)$$

Since this is a SLS calculation, no load factors are used; therefore,

$$\Delta = \frac{5 \times 0.28 \times 6000^4}{384 \times 210000 \times 11.28 \times 10^6} + \frac{10 \times 10^3 \times 6000^3}{48 \times 210000 \times 11.28 \times 10^6} = 2.0 + 19.0 = 21 \text{ mm}$$

The maximum allowable deflection is span/200 = 30 mm; therefore, the beam is satisfactory.

2.2.1 Bending moment capacity in the presence of high shear forces

Von Mises yield criteria (Equation 2.1) show us that shear stresses reduce the tensile or compressive stress required to cause yielding. In fact, Equation 2.1 shows that the tensile strength falls to zero if the applied shear stress equals the yield shear stress. Therefore, for beams that resist high shear forces combined with high moments, the bending strength should be reduced.

The flanges of hot-rolled sections develop low shear stresses, as shown by the shear stress distribution sketched in Figure 2.2a. Therefore, the effects of shear stresses on the tensile strength of the flange material can be safely ignored.

Von Mises yield criteria can be used to determine the moment capacity in the presence of high shear. It will be seen from the following worked example that the bending strength of hot-rolled I-sections is not greatly influenced by high shear stresses. For this reason, no reduction in moment capacity is necessary unless the applied shear force is greater than 50% of the shear strength. Most hot-rolled sections have relatively high shear strengths, because the webs are made thick to prevent distortion when cooling after rolling. Therefore, in most practical situations, the moment capacity of hot-rolled sections is unaffected by shear.

Example 2.5: Bending moments combined with high shear forces

A Class 1 I-section beam has a depth of 260 mm, web thickness of 6.3 mm, plastic section modulus of 353 cm^3 and yield stress of 355 N/mm^2.

1. Determine the bending strength, $M_{pl, Rd}$.
2. Determine the shear strength, $V_{pl, Rd}$.
3. Determine bending moment capacity if the applied shear force is equal to the shear strength, i.e., $V_{Ed} = V_{pl, Rd}$.
4. Determine bending moment capacity if $V_{Ed} = 0.75 V_{pl, Rd}$.

1. From Equation 2.9, the bending strength is

$$M_{pl, Rd} = 355 \times 353 \times 10^3 \times 10^{-6} = 125.3 \text{ kN.m}$$

2. From Equation 2.4, the shear area is

$$A_v = 6.3 \times 260 = 1638 \text{ mm}^2$$

From Equation 2.3, the shear strength is

$$V_{pl, Rd} = \frac{1638 \times 355}{\sqrt{3}} \times 10^{-3} = 336 \text{ kN}$$

3. The Von Mises equation (Equation 2.1) shows us that the shear area cannot resist bending stresses, because the applied shear stress is equal to the yield shear stress (Equation 2.2). The moment capacity of the shear area (i.e. the web) is therefore zero, although the moment capacity of the flanges is unchanged, because the flanges do not contribute significantly to the shear strength. Therefore, the moment capacity of the shear area is calculated and

deducted from the moment capacity of the section (125.3 kN.m). The plastic section modulus of a rectangular block of width t and depth D is

$$W_{pl} = \frac{t \times D^2}{4} \tag{2.15}$$

Therefore, the plastic section modulus of the shear area is

$$W_{pl,\,web} = \frac{6.3 \times 260^2}{4} = 106470 \text{ mm}^3$$

and the bending strength of the shear area in the absence of shear stresses is

$$M_{web} = 355 \times 106470 \times 10^{-6} = 37.8 \text{ kN.m}$$

Therefore, the reduced bending strength is the full bending strength minus the web bending strength, i.e.,

$$M_{pl,\,Rd} = 125.3 - 37.8 = 87.5 \text{ kN.m}$$

4. The applied shear force is 3/4 of the shear strength. From Equation 2.2, it follows that the applied shear stress is

$$\tau_{xy} = \frac{3}{4} \times \tau_y = \frac{3}{4} \times \frac{355}{\sqrt{3}} = 154 \text{ N/mm}^2$$

and Equation 2.1 becomes

$$\sqrt{\sigma_x^2 + 3\tau_{xz}^2} \leq f_y$$

$$\sqrt{\sigma_x^2 + 3 \times 154^2} = 355$$

$$\sigma_x = 234 \text{ N/mm}^2$$

This means that the tensile strength of the shear area has fallen from 355 N/mm² to 234 N/mm² due to the applied shear stress. The reduction in moment capacity of the shear area is

$$M_{reduction} = (355 - 234) \times 106470 \times 10^{-6} = 12.9 \text{ kN.m}$$

Thus, the reduced moment capacity is

$$M_{pl,\,Rd} = 125.3 - 12.9 = 112.4 \text{ kN.m}$$

In summary, a shear force equal to the shear strength reduced the bending strength by 30%, whereas a shear of 75% of the shear strength reduced the bending strength by only 10%.

2.3 LATERAL TORSIONAL BUCKLING

During bending, one half of a beam is thrown into compression, and this can cause buckling in a similar manner to the buckling of a strut. This is known as *lateral torsional buckling* or *LTB* and it is illustrated in Figure 2.9. Unlike a strut, half the beam will be in tension, and the tensile force will help to restrain buckling. This restraint will cause the beam to twist as it buckles and the beam's torsional stiffness will resist this twisting. Tubular members have a high torsional stiffness and therefore do not normally experience LTB. However, I-section beams have very low torsional stiffness and are therefore highly susceptible.

LTB is responsible for a large proportion of collapses of steel-framed structures. It is a particular problem during construction, when the steel may not be fully restrained against sideways movement. Site workers have been killed when temporary restraint against LTB has not been installed. Since this problem is not always obvious, it is the responsibility of the designer to communicate to the construction team the need for temporary restraints.

The main factors that affect LTB are as follows:

1. *Lateral restraint.* If the compression flange is restrained against sideways movement, then LTB will not occur. It is important to appreciate that (a) *restraining the tension flange will not prevent LTB* and (b) the *compression flange is not always the top flange.*
2. *Torsional stiffness.* Open cross sections, like I- and H-sections, have low torsional stiffness and therefore have little ability to resist twisting. Conversely, closed sections (hollow sections) have high torsional stiffness and are much less prone to LTB.
3. Beams in which the major axis second moment of area are much greater than the minor axis second moment of area are particularly vulnerable to LTB, i.e., I-sections.

Elastic critical buckling moment. For a beam, the relationship between the torsional moment (T) and the angle of twist (θ) is

$$T = GI_t \frac{d\theta}{dx} \tag{2.16}$$

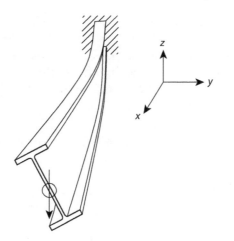

Figure 2.9 Lateral torsional buckling caused by vertical loading to a cantilever (note the twisting).

where
  x is position along the beam.
  G is the shear modulus.
  I_t is the torsional constant.

The product of GI_t is known as the *torsional stiffness*. For open cross sections, like I- and H-sections, an extra term to account for warping is included (see Figure 2.10) and the equation becomes

$$T = GI_t \frac{d\theta}{dx} - EI_w \frac{d^3\theta}{dx^3} \qquad (2.17)$$

where
  I_w is the warping constant.

During LTB, deformation occurs about the x-, y- and z-axes and these deformations are interrelated in the form of three simultaneous differential equations, the solution of which is known as the *elastic critical buckling moment*, given as

$$M_{cr} = C \frac{\pi}{L_{cr}} \sqrt{\frac{EI_z}{1 - I_z / I_y} \left(GI_t + \frac{EI_w \pi^2}{L_{cr}^2} \right)} \qquad (2.18)$$

where
  I_y is the major axis second moment of area.
  I_z is the minor axis second moment of area.
  L_{cr} is the effective length.
  C is the equivalent uniform moment factor (see Figure 2.11).

Simplification of the M_{cr} formula. In most practical situations, C makes little difference to strength. Therefore, in the interests of simplicity, it can be set as equal to 1.0 for all end conditions and therefore eliminated from the design process. In addition, the resistance to warping at the ends of the beam can also be neglected with only a slight loss of efficiency. These changes lead to the following simpler expression:

$$M_{cr} = \frac{\pi}{L_{cr}} \sqrt{\frac{EI_z GI_t}{1 - I_z/I_y}} \qquad (2.19)$$

Effective Length, L_{cr}. It is vital to use the correct effective length when designing laterally unrestrained beams, since this critically affects the load capacity. Effective length is defined using a very similar approach to that for struts (see Chapter 3), with the

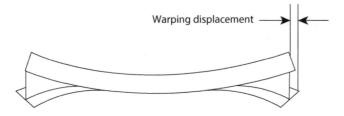

Figure 2.10 Warping at the ends of a beam due to twisting.

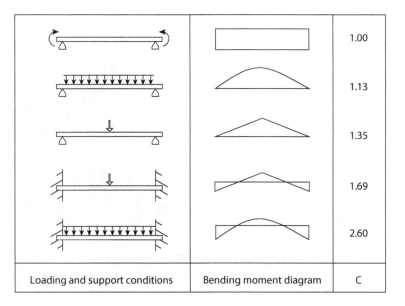

Loading and support conditions	Bending moment diagram	C
		1.00
		1.13
		1.35
		1.69
		2.60

Figure 2.11 Equivalent uniform moment factor, C.

objective to determine the half sine wave buckling mode. Important factors to consider include

- *Lateral restraint.* The effective length is based on the distance between lateral restraints. *Providing lateral restraint to the tension flange will not prevent LTB from occurring.* Therefore, lateral restraints must be provided to the centre of the beam or to the compression flange.
- *Destabilising loads.* Loads that are supported by the top flange *and* free to displace sideways are destabilising. This type should be avoided, because the load can develop a torsional moment (see Figure 2.12). If destabilising loads cannot be avoided, then the effective length should be increased by 20%.
- *Support conditions.* The degree of torsional restraint provided by the supports has a critical influence on LTB. Connections that would be considered as torsionally restrained include end plate connections, as shown in Figure 2.13. The fin plate connections can be too flexible and may not have sufficient stiffness to resist LTB. If beams are not torsionally restrained at the supports, then the bending strength will be adversely affected; therefore, the effective length should be increased.

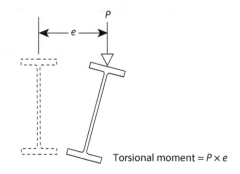

Figure 2.12 Lateral movement due to a destabilising load.

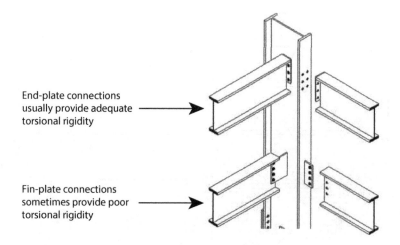

End-plate connections usually provide adequate torsional rigidity

Fin-plate connections sometimes provide poor torsional rigidity

Figure 2.13 Connections with different torsional stiffness.

The calculation of effective length is an important concept and is best described using some commonly occurring examples:

Case 1. Consider a beam that supports the wet weight of concrete along its length and that has partial depth end plate connections as shown in Figure 2.13 (these provide adequate torsional restraint). If the load is applied to the top flange and the load is free to move sideways, then the wet concrete is classified as destabilising and the effective $L_{cr} = 1.2L$, where L is the distance between supports.

Case 2. Consider a primary beam that supports secondary beams at third span points. These beams apply their loads at the centre of the beam and are not therefore destabilising. If the secondary beams are capable of provide bracing, then $L_{cr} = L/3$.

Case 3. Consider the common case of a beam supporting an opening in a masonry wall. The loading from the masonry wall is applied to the top flange and should therefore be considered as destabilising. In addition, the masonry bearings (padstones) will provide no torsional restraint. This adverse combination of factors requires an increased effective length, where $L_{cr} = 1.2L + 2h$, where h is the beam depth and L is the span.

Design moment. The elastic critical moment, M_{cr}, is the theoretical upper limit for bending strength. M_{cr} is by definition purely elastic and can far exceed the yield moment; therefore, it needs to be capped for design purposes. The region of the graph shown in Figure 2.14 bounded by M_{cr} and $M_{pl,Rd}$ represents the theoretical upper limit on bending strength.

Imperfections will reduce the strength well below this theoretical upper limit. The most important imperfections are the internal shrinkage stresses caused by welding or hot-rolling. These throw the flange tips of I- and H-sections into compression, and this leads to a significant reduction in lateral torsional buckling strength. A quick estimate of the design moment shown as $M_{b,Rd}$ in Figure 2.14 can be obtained using a Gordon–Rankine (empirical) approximation, as follows:

$$\frac{1}{M_{b,Rd}} = \frac{1}{M_{pl,Rd}} + \frac{1}{M_{cr}} \tag{2.20}$$

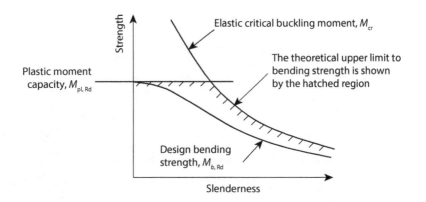

Figure 2.14 Relationship between moment capacity and slenderness.

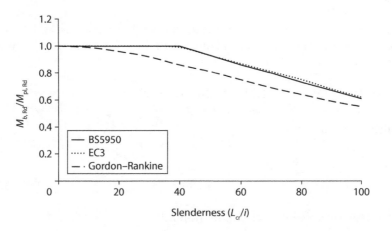

Figure 2.15 Comparison between the simplified method and the code-based methods.

or

$$M_{b,\,Rd} = \left(\frac{1}{M_{pl,\,Rd}} + \frac{1}{M_{cr}} \right)^{-1} \qquad (2.21)$$

As with any buckling problem, members become increasingly vulnerable to imperfections or sideways forces as slenderness increases. For this reason, it is not advisable to use slender beams. Figure 2.15 shows a comparison between the moment capacities calculated using this Gordon–Rankine type approximation and the full code-based methods that take account of imperfections more formally.

Example 2.6: Lateral torsional buckling check for bridge girders supporting wet concrete

The concrete slab of a bridge deck is supported by I-section beams as shown in Figure 2.16. The beams span 8.5 m between supports that provide torsional restraint to the ends of the beams. During casting of the slab, an imposed load of 1.0 kN/m² is applied to account for the weight of workers and equipment. Determine if the beams can support this load safely.

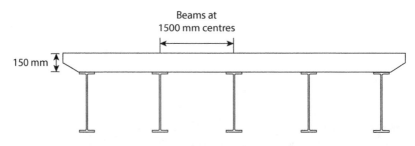

Figure 2.16 Cross section through a bridge deck.

The beams have the following properties: f_y = 275 N/mm², E = 210,000 N/mm², G = 80,770 N/mm², $W_{pl,y}$ = 1501 cm³, I_y = 21,370 cm⁴, I_z = 1545 cm⁴, I_t = 62.8 cm⁴, beam self-weight = 0.726 kN/m.

The beams are spaced 1.5 m apart; therefore, each inner beam supports a 1.5 m wide section of slab. Assuming a density of reinforced concrete of 25 kN/m³ and using Equation 1.3, the ULS (factored) dead + imposed load per beam is

$$w = 1.35 \times 25 \times 0.150 \times 1.5 + 1.35 \times 0.726 + 1.5 \times 1.0 \times 1.5 = 10.8 \text{ kN/m}$$

and the applied moment is

$$M_{Ed} = \frac{wL^2}{8} = \frac{10.8 \times 8.5^2}{8} = 97.5 \text{ kN.m}$$

The plastic moment of resistance from Equation 2.9 is

$$M_{pl,Rd} = 275 \times 1501 \times 10^3 \times 10^{-6} = 413 \text{ kN.m}$$

The applied moment is therefore less than 1/4 of the beam plastic moment capacity. However, the beams are free to buckle sideways, because wet concrete does not provide sideways resistance. The beams are torsionally restrained at the supports, and under these conditions the effective length would normally be 1.0 L. However, the load is applied to the top flange and is therefore classed as a *destabilising load*. The effective length is therefore increased by 20%:

$$L_{cr} = 1.2 \times L = 1.2 \times 8500 = 10200 \text{ mm}$$

The elastic critical buckling moment from Equation 2.19 is

$$M_{cr} = \frac{\pi}{L_{cr}} \sqrt{\frac{EI_z GI_t}{1 - I_z / I_y}}$$

$$M_{cr} = \frac{\pi}{10200} \times 10^{-6} \sqrt{\frac{210000 \times 1545 \times 10^4 \times 80770 \times 62.8 \times 10^4}{1 - 1545 / 21370}} = 130 \text{ kN.m}$$

and the bending strength from Equation 2.21 is

$$M_{b,\,Rd} = \left(\frac{1}{M_{pl,\,Rd}} + \frac{1}{M_{cr}} \right)^{-1}$$

$$M_{b,\,Rd} = \left(\frac{1}{413} + \frac{1}{130} \right)^{-1} = 99 \text{ kN.m}$$

Since the capacity of 99 kN.m is greater than the applied moment of 97.5 kN.m, the beams should in theory be sufficiently strong. However, since the beam is very slender ($M_{b,\,Rd} \ll M_{pl,\,Rd}$), the beam will be vulnerable to impacts or imperfections. Engineers term this a lack of *robustness*. For example, an accidental impact could trigger failure or overloading may occur due to concrete being initially piled up at midspan. To guard against either of these eventualities, it would be prudent to install temporary bracing to raise the buckling moment and eliminate the danger.

Example 2.7: Beam design involving LTB

A bridge comprises two I-section beams that span 8 m between simple supports and support a 6 m wide, 220 mm deep reinforced concrete deck.

1. Determine the LTB moment capacity of a beam, if the beams are torsionally restrained at the supports but laterally unrestrained along their lengths.
2. The beams are laterally unrestrained when resisting the wet weight of the concrete in addition to an imposed load of 0.75 kN/m² to account for the weight of the construction staff and plant. Determine if the beams need temporary restraints against lateral torsional buckling during construction.

Beam properties
$f_y = 275 \text{ N/mm}^2$, $E = 210,000 \text{ N/mm}^2$, $G = 80,770 \text{ N/mm}^2$, depth = 500 mm, web thickness = 10 mm, $W_{pl,\,y} = 1470 \text{ cm}^3$, $I_y = 29,400 \text{ cm}^4$, $I_z = 1450 \text{ cm}^4$, $I_w = 0.70 \times 10^{12} \text{ mm}^6$, $I_t = 37.1 \text{ cm}^4$, self-weight = 70 kg/m

1. From Equation 2.9, the plastic moment capacity is

$$M_{pl,\,Rd} = 275 \times 1470 \times 10^3 \times 10^{-6} = 404.3 \text{ kN.m}$$

The more exact method for calculating the elastic critical moment will be used in this example. From Figure 2.11, the equivalent uniform moment factor (C) for a simply supported beam supporting a UDL is 1.13. In addition, the load is applied to the top flange and is therefore a *destabilising load* and the effective length is increased by 20%; therefore,

$$L_{cr} = 1.2 \times L$$

$$L_{cr} = 1.2 \times 8000 = 9600 \text{ mm}$$

Using Equation 2.18

$$M_{cr} = C \frac{\pi}{L_{cr}} \sqrt{\frac{EI_z}{1 - I_z / I_y} \left(GI_t + \frac{EI_w \pi^2}{L_{cr}^2} \right)}$$

$$M_{cr} = 1.13 \times \frac{\pi}{9600}$$

$$\times 10^{-6} \sqrt{\frac{210000 \times 1450 \times 10^4}{1 - 1450 / 29400} \left(80770 \times 37.1 \times 10^4 + \frac{210000 \times 0.70 \times 10^{12} \pi^2}{9600^2} \right)}$$

$$= 141.5 \text{ kN.m}$$

From Equation 2.21, the moment capacity is

$$M_{b, Rd} = \left(\frac{1}{M_{pl, Rd}} + \frac{1}{M_{cr}} \right)^{-1} = \left(\frac{1}{404.3} + \frac{1}{141.5} \right)^{-1} = 104.8 \text{ kN.m}$$

2. The two beams each support half of the 6 m wide bridge. The ULS uniformly distributed load per beam of the wet concrete, steel sections, construction workers and equipment is

$$w = 1.35 \times (25 \times 3 \times 0.22 + 9.81 \times 70 \times 10^{-3}) + 1.5 \times 3 \times 0.75 = 26.6 \text{ kN/m}$$

$$M_{Ed} = \frac{26.6 \times 8^2}{8} = 212.8 \text{ kN.m}$$

The applied moment (212.8 kN.m) is less than the plastic moment capacity (404.3 kN.m). However, it is much greater than the LTB moment capacity (104.8 kN.m); therefore, the bridge will collapse. Since buckling occurs without warning, this would probably cause casualties. The bridge should therefore be braced against sideways movements during construction.

Example 2.8: Beam design for a multistorey building

Figure 2.17 shows the framing arrangement for the floor of a multistorey steel-framed build-ing. The frame is 'simple', i.e., it is assumed the joints between the beams and columns are effectively pinned. The beams support precast concrete slabs that sit on the top flanges of the steel sections, with the direction of span indicated by the arrows. The unfactored dead load (inclusive of beam self-weight) = 7 kN/m² and the unfactored imposed load = 3.5 kN/m², f_y = 265 N/mm², E = 210,000 N/mm², G = 80,770 N/mm².

1. Check the shear strength, bending strength and deflection of Beam A, which is laterally restrained.
2. Check the bending strength of Beam B if it is laterally restrained by the supported beams but laterally unrestrained between the loading points.

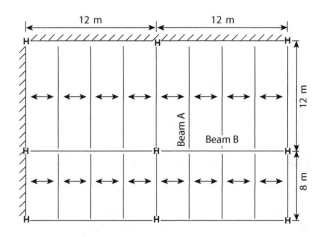

Figure 2.17 Plan showing framing arrangement.

Beam properties
Beam A: I_y = 126,000 cm⁴, $W_{pl, y}$ = 4590 cm³, web thickness = 11.8 mm, section depth = 612.4 mm

Beam B: $W_{pl, y}$ = 17,700 cm³, I_y = 720,000 cm⁴, I_z = 45,400 cm⁴, I_t = 1730 cm⁴

1. *Beam A – Laterally restrained*. This is known as a *secondary beam* and these are spaced at 3 m centres; therefore, it supports a 3 m wide section of floor slab and the ultimate limit state UDL from Equation 1.3 is

$$w = 3.0 \times (1.35 \times 7 + 1.5 \times 3.5) = 44.1 \text{ kN/m}$$

This produces a support shear force:

$$V_{Ed} = \frac{44.1 \times 12}{2} = 264.6 \text{ kN}$$

and the shear strength from Equation 2.3 is

$$V_{pl, Rd} = \frac{11.8 \times 612.4 \times 265}{\sqrt{3}} \times 10^{-3} = 1106 \text{ kN} \gg V_{Ed} \therefore \text{pass}$$

The applied midspan moment is

$$M_{Ed} = \frac{44.1 \times 12^2}{8} = 793.8 \text{ kN.m}$$

Beam A is considered as laterally restrained due to the frictional force between the concrete slabs and the top flange of the beams. The moment capacity is therefore safely calculated using Equation 2.9:

$$M_{pl, Rd} = f_y W_{pl}$$

$$M_{pl, Rd} = 4590 \times 10^3 \times 265 \times 10^{-6} = 1216 \text{ kN.m} > M_{Ed} \therefore \text{pass}$$

The beam deflection must now be checked. The serviceability limit state design load from Equation 1.2 is

$$w = 3.0 \times (7 + 3.5) = 31.5 \text{ kN/m} = 31.5 \text{ N/mm}$$

and the midspan deflection from Equation 1.6 is

$$\delta = \frac{5 \times 31.5 \times 12000^4}{384 \times 210000 \times 126000 \times 10^4} = 32 \text{ mm}$$

A conventional limit on deflection is span/360, which would provide a limit of 33 mm; therefore, this beam is sufficiently stiff.

2. *Beam B – Laterally unrestrained.* The analysis in point (1) shows that that the load on the beams supported by Beam B is 44 kN/m. This will produce point loads, P, at quarter span points, where

$$P = 44.1(12 + 8)/2 = 441 \text{ kN}$$

The midspan moment for a beam supporting point loads (P) at quarter span points is

$$M = \frac{PL}{2} \tag{2.22}$$

Therefore,

$$M_{Ed} = \frac{441 \times 12}{2} = 2646 \text{ kN.m}$$

The basic plastic moment of resistance is

$$M_{pl, Rd} = 17700 \times 10^3 \times 265 \times 10^{-6} = 4691 \text{ kN.m}$$

The beam is laterally restrained by the supported beams but laterally unrestrained in between. Therefore, the effective length is 3000 mm (span/4), and from Equation 2.19 the elastic critical moment, is

$$M_{cr} = \frac{\pi}{L_{cr}} \sqrt{\frac{EI_z GI_t}{1 - I_z/I_y}}$$

$$M_{cr} = \frac{\pi}{3000} \times 10^{-6} \sqrt{\frac{210000 \times 45400 \times 10^4 \times 80770 \times 1730 \times 10^4}{1 - 45400/720000}} = 12487 \text{ kN.m}$$

and the buckling moment from Equation 2.20 is

$$M_{b, Rd} = \left(\frac{1}{M_{pl, Rd}} + \frac{1}{M_{cr}} \right)^{-1}$$

$$M_{b, Rd} = \left(\frac{1}{12487} + \frac{1}{4691} \right)^{-1} = 3410 \text{ kN.m} > M_{Ed} \therefore \text{OK}$$

Problems

Solutions to these problems are provided at https://www.crcpress.com/9781498741217

P.2.1. A simply supported beam spans 5 m and supports an unfactored UDL of 100 kN/m dead (including self-weight) and 200 kN/m imposed.
   a. Determine the ULS design load per m.
   b. Determine the ULS design moment.
   c. Determine the minimum plastic section modulus required to support these loads if the yield stress is 275 N/mm².
   d. Determine the ULS shear force at the supports.
   e. Determine the minimum shear area required to resist this shear force.
   Ans. (a) 435 kN/m, (b) 1359 kN/m, (c) 4942 cm³, (d) 1087.5 kN and (e) 6849 mm².

P.2.2. Figure 2.18 shows a hot-rolled I-section beam (section classification is Class 1).

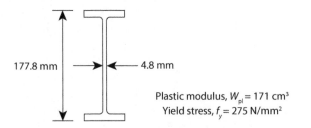

177.8 mm ← → 4.8 mm

Plastic modulus, W_{pl} = 171 cm³
Yield stress, f_y = 275 N/mm²

Figure 2.18 Section details.

   a. Determine the design moment capacity, $M_{pl, Rd}$.
   b. Determine the design shear capacity, $V_{pl, Rd}$.
   c. Prove that this section will not fail by shear buckling.
   d. Determine $M_{pl, Rd}$ if the applied shear force = 100% of $V_{pl, Rd}$.
   e. Determine $M_{pl, Rd}$ if the applied shear force = 75% of $V_{pl, Rd}$.
   Ans. (a) 47.0 kN.m, (b) 135.3 kN, (c) τ_{cr} (739 N/mm²) > τ_y (158.8 N/mm²), (d) 36.6 kN.m and (e) 43.5 kN.m.

P.2.3. An I-section beam resists end moments, as shown in Figure 2.19.

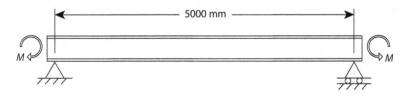

Figure 2.19 Beam with end moments.

   a. Determine the plastic moment of resistance if $W_{pl, y}$ = 393 cm³ and f_y = 275 N/mm².
   b. Determine the effective length if the beam is laterally unrestrained along its length but torsionally restrained at the supports.
   c. Determine the elastic critical buckling moment if Young's modulus, E = 210,000 N/mm², G = 80,770 N/mm², I_y = 4413 cm⁴, I_z = 448 cm⁴ and I_t = 8.55 cm⁴.
   d. Determine the lateral torsional buckling design moment.
   Ans. (a) 108 kN.m, (b) 5000 mm, (c) 53.4 kN.m and (d) 35.7 kN.m.

Chapter 3

Buckling of steel columns and trusses

This chapter explains how to estimate the buckling capacity of steel members subjected to compression and bending. The method used is a Gordon–Rankine approach, which provides a quick estimate of strength that is slightly conservative in comparison with the full code-based methods. The equations presented in this chapter do not include a partial safety factor on materials. This is because the partial safety factor for steelwork member design is taken as 1.0 in the Eurocodes.

3.1 BASIC STRUT BUCKLING

A strut is an axially loaded member under pure compression. This is different from the columns in buildings, which are also subjected to bending moments and are therefore called *beam columns*. Petrus van Musschenbroek discovered in 1729 that the elastic critical buckling force (N_{cr}) is inversely proportional to the length squared, i.e.,

$$N_{cr} \propto \frac{1}{L^2}$$

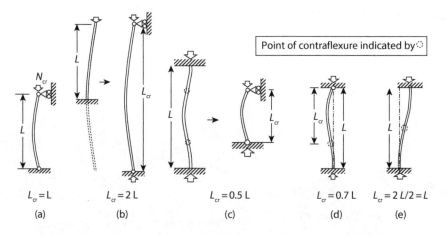

Point of contraflexure indicated by ☉

$L_{cr} = L$ $L_{cr} = 2L$ $L_{cr} = 0.5L$ $L_{cr} = 0.7L$ $L_{cr} = 2L/2 = L$

(a)   (b)   (c)   (d)   (e)

Figure 3.1 The concept of effective length for differing end conditions.

Leonard Euler (1757) subsequently showed that for a pin-ended strut (Figure 3.1a)

$$N_{cr} = \frac{\pi^2 S}{L^2}$$

Euler also solved the differential equations for struts with a variety of different end conditions. For example, he showed that the critical buckling load for a strut fixed at the base and free to sway at the tip (Figure 3.1b) is

$$N_{cr} = \frac{\pi^2 S}{4L^2}$$

The term S was later defined by Claude-Louis Navier in 1826 as the bending stiffness, EI, where I is the second moment of area and E is Young's modulus.

Effective length, L_{cr}. It is possible to derive separate equations for struts with a range of different end conditions. However, for design purposes, *the same equation is used throughout*, with the length L replaced by effective length, L_{cr}, which compensates for the different end conditions.

$$N_{cr} = \frac{\pi^2 EI}{L_{cr}^2} \tag{3.1}$$

The effective length is the length of an equivalent strut with pinned end conditions (Figure 3.1a). For the cantilever strut shown in Figure 3.1b, the equivalent pin-ended strut is not obvious. It is necessary to project the strut into the support to provide the equivalent pin-ended strut, where L_{cr} is twice the length of the original strut. It is sometimes necessary to identify *points of contraflexure*. These are the point(s) of zero moment. Since the moment at these points is zero, the response of the strut would be unchanged if a hinge was introduced at that exact location. This is why points of contraflexure are used when determining L_{cr}. In the strut shown in Figure 3.1c, points of contraflexure occur at the quarter points; thus, $L_{cr} = 0.5 L$. Figure 3.1d and e shows the points of contraflexure for some other commonly occurring end conditions.

Slenderness. The technical definition of slenderness (λ) is

$$\lambda = \frac{L_{cr}}{i}$$

where i is the radius of gyration. Since the second moment of area, $I = Ai^2$, the Euler critical buckling equation can be reconfigured in terms of slenderness:

$$N_{cr} = \frac{\pi^2 EA}{\lambda^2} \tag{3.2}$$

Slenderness is a useful design term, because it combines the parameters of length and radius of gyration.

Inelastic bucking. Euler's formula applies only to struts that remain elastic during buckling, i.e., ones that return to their original position when the load is removed. This occurs only in very slender struts. In practice, columns fail by a combination of buckling and yielding, in which case Euler's formula will overpredict strength. Figure 3.2 shows the relationship between slenderness and the elastic critical buckling force. The actual strength will not be greater than the crushing strength, which is given by

$$N_{pl, Rd} = f_y \times \text{Area} \tag{3.3}$$

Therefore, the theoretical upper limit on strength is the lesser of either N_{cr} or $N_{pl, Rd}$ and is shown as the hatched region on Figure 3.2. The actual strength will be somewhat below this region because of imperfections, and the design buckling force ($N_{b, Rd}$) can be estimated using the (empirical) Gordon–Rankine formula, which was routinely used from 1862 onwards after appearing in Rankine's famous book entitled *A Manual of Civil Engineering.* In Rankine's method

$$\frac{1}{N_{b, Rd}} = \frac{1}{N_{pl, Rd}} + \frac{1}{N_{cr}}$$

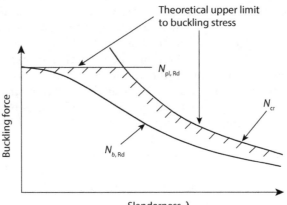

Figure 3.2 Relationship between slenderness and buckling stress of struts.

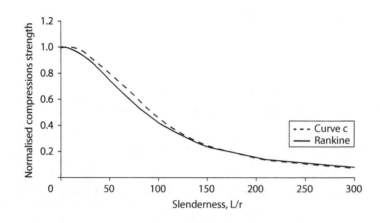

Figure 3.3 Comparison between Rankine–Gordon method and Eurocode 3 for H-sections buckling about the weak axis (illustrated in Figure 3.4c).

which rearranges to

$$N_{b,\text{Rd}} = \left(\frac{1}{N_{\text{pl,Rd}}} + \frac{1}{N_{\text{cr}}} \right)^{-1} \tag{3.4}$$

This formula works because $N_{b,\text{Rd}} \rightarrow N_{\text{pl,Rd}}$ when slenderness is small and $N_{b,\text{Rd}} \rightarrow N_{\text{cr}}$ when slenderness is high. Modern codes use semi-empirical approaches to define $N_{b,\text{Rd}}$ and Figure 3.3 shows the Eurocode 3 design strength compared with the Gordon–Rankine strength. For the most important condition of H-sections buckling about the weak axis (curve c), this shows that this Gordon–Rankine method gives a good level of accuracy. In this book, Equation 3.4 has been adapted to analyse lateral torsional buckling, arch buckling and plate buckling, because this technique is simple to understand and generally very accurate.

The effect of residual stresses. Most steel sections are formed using a hot rolling process, whereby steel is softened by heating in a furnace. The parts of the section that cool fastest end up in permanent compression, which is internally balanced by tension in the parts that cool slowest. The resulting internal stresses are known as *residual* or *shrinkage stresses*. Similar stress patterns are caused when plates are welded to form sections.

Residual stresses are safely ignored when calculating the elastic or plastic moment capacities. However, they are important when considering buckling or fatigue. Residual stresses also cause members to bend that are split along their lengths; for example, when T-sections are formed by splitting H-sections down the middle. Residual stresses can commonly be as high as 50% of the yield stress and they will lower the buckling strength. This is because residual stresses promote early yielding and yielding often initiates buckling.

Figure 3.4a shows the residual stress distribution in a hot-rolled H-section. If prevented from buckling about the minor axis (Figure 3.4b), then the adverse effects of the compression residual stresses are counterbalanced by the positive effect of the tensile residual stresses. The more common case is (Figure 3.4c) minor axis buckling. In this case, the maximum buckling stress (at the tip of the flanges) coincides with the position of maximum residual compression stress. This is the worst combination and causes early buckling. If the section uses very thick steel, the residual stresses will be higher and sections built up from plate using welding have very high residual stresses and hence the lowest permitted design stresses.

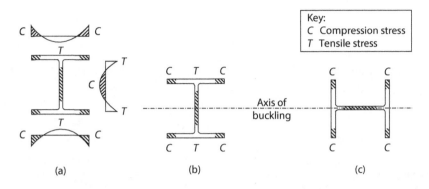

Figure 3.4 Residual stresses for hot-rolled H-sections. (a) Residual stress distribution in web and flange, (b) major axis buckling and (c) minor axis buckling.

Example 3.1: Strut design

A rectangular hollow section strut is 2.5 m long and is pinned at its supports. The yield stress is 355 N/mm², $E = 210,000$ N/mm², $I_y = 607$ cm⁴, $I_z = 324$ cm⁴ and $A = 19.2$ cm².

1. Estimate the design compression strength if the strut is free to buckle about either the y–y or z–z axes.
2. Estimate the buckling strength if buckling about the weak axis (z–z axis) is prevented by bracing.

1. The first step is to determine the crushing strength using Equation 3.3:

$$N_{pl, Rd} = f_y \times \text{Area}$$

$$N_{pl, Rd} = 355 \times 19.2 \times 10^2 \times 10^{-3} = 682 \text{ kN}$$

The strut is pinned at the supports; therefore, the buckling mode is identical to that sketched in Figure 3.1a and the effective length is

$$L_{cr} = 1.0 \times L = 2500 \text{ mm}$$

The strut will buckle about the weakest axis unless prevented from doing otherwise. Therefore, the weak axis second moment of area (I_z) is used in the elastic critical buckling force calculation using Equation 3.1:

$$N_{cr} = \frac{\pi^2 E A}{L_{cr}^2}$$

$$N_{cr} = \frac{\pi^2 \times 210000 \times 324 \times 10^4}{2500^2} \times 10^{-3} = 1074 \text{ kN}$$

and using the Gordon–Rankine approximation, Equation 3.4, the design strength is

$$N_{b, Rd} = \left(\frac{1}{N_{pl, Rd}} + \frac{1}{N_{cr}} \right)^{-1}$$

$$N_{b,\,\mathrm{Rd}} = \left(\frac{1}{682} + \frac{1}{1074}\right)^{-1} = 417 \text{ kN}$$

2. Since buckling about the weak (z–z) axis is prevented, the major axis second moment of area (I_y) is used in Equation 3.1:

$$N_{\mathrm{cr}} = \frac{\pi^2 \times 210000 \times 607 \times 10^4}{2500^2} \times 10^{-3} = 2013 \text{ kN}$$

and from Equation 3.4,

$$N_{b,\,\mathrm{Rd}} = \left(\frac{1}{682} + \frac{1}{2013}\right)^{-1} = 509 \text{ kN}$$

There is a significant difference between the buckling strength between (1) and (2) and this highlights the importance of correctly identifying which of I_y and I_z should be used when calculating the elastic critical buckling force.

3.2 BEAM COLUMNS

Members subjected to bending combined with axial compression need to be checked to ensure that they will not buckle. The first check is to ensure that the section remains elastic under the applied loading. Basic theory tells us that the maximum stress in a section subjected to an axial force and biaxial moments is

$$\frac{N}{A} + \frac{M_y}{W_{\mathrm{el},\,y}} + \frac{M_z}{W_{\mathrm{el},\,z}} = \sigma_{\max}$$

where
  $\sigma_{\max}$ is the maximum stress in the section.
  N is the applied axial load.
  M_y and M_z are the applied moments about the strong (y–y) and weak (z–z) axes, respectively.
  $W_{\mathrm{el},\,y}$ and $W_{\mathrm{el},\,z}$ are the elastic section moduli.

Dividing this through by the yield stress, f_y, gives

$$\frac{N}{A f_y} + \frac{M_y}{W_{\mathrm{el},\,y} f_y} + \frac{M_z}{W_{\mathrm{el},\,z} f_y} = \frac{\sigma_{\max}}{f_y}$$

The maximum stress must not exceed the yield stress; therefore,

$$\frac{N}{N_{\mathrm{pl},\,\mathrm{Rd}}} + \frac{M_y}{M_{\mathrm{el},\,y}} + \frac{M_z}{M_{\mathrm{el},\,z}} \leq 1$$

This equation is useful because it quantifies the degree of utilisation before yielding (ignoring residual stresses), although it does not include a reduction in strength due to buckling.

In order to check for buckling, the crushing force ($N_{pl,\,Rd}$) must be replaced by the buckling force ($N_{b,\,Rd}$) defined in Equation 3.4, i.e.,

$$\frac{N}{N_{b,\,Rd}} + \frac{M_y}{M_{el,\,y}} + \frac{M_z}{M_{el,\,z}} \leq 1$$

This equation is incomplete, because the axial force amplifies moments. The applied end moments are fixed in magnitude, but the moment part way down the strut is magnified and it is the moment in the middle that creates buckling. The increased moment increases deflections and this will be self-propagating if above a critical value. This process is illustrated in Figure 3.5a, which shows a column subjected to end moments.

Figure 3.5b shows the additional moment due to the axial force multiplied by the deflection induced by the end moments ($N.\delta$). It was shown by Timoshenko and Gere (1961) that the axial force will amplify the end moment by the following factor:

$$\alpha = \frac{1}{1 - N / N_{cr}} \tag{3.5}$$

This 'amplification factor' is only approximate, but it is still very useful and it appears in many different design formulae. It sometimes appears as

$$\frac{N_{cr}}{N_{cr} - N}$$

The final beam–column design equation becomes

$$\frac{N}{N_{b,\,Rd}} + \frac{\alpha_y M_y}{M_{el,\,y}} + \frac{\alpha_z M_z}{M_{el,\,z}} \leq 1 \tag{3.6}$$

where
α_y and α_z are the moment amplification factors for the major and minor axes, respectively.
M_y and M_z are applied moments about the major and minor axes, respectively.
$M_{el,\,y}$ and $M_{el,\,z}$ are the major and minor axis elastic moment capacities, respectively.

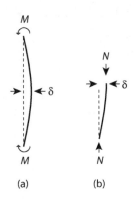

(a)   (b)

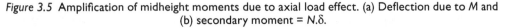

Figure 3.5 Amplification of midheight moments due to axial load effect. (a) Deflection due to *M* and (b) secondary moment = *N*.δ.

The additional (secondary) moments are only important for slender members. They are not significant for 'stocky' members, such as the columns found in most multistorey office buildings, in which case the interaction equation shortens to

$$\frac{N}{N_{b,\,Rd}} + \frac{M_y}{M_{el,\,y}} + \frac{M_z}{M_{el,\,z}} \leq 1 \tag{3.7}$$

Lateral torsional buckling. If a beam column is susceptible to lateral torsional buckling, as described in Section 2.3, then the lateral torsional buckling design moment, $M_{b,\,Rd}$ (Section 2.3), replaces $M_{el,\,y}$, i.e.,

$$\frac{N}{N_{b,\,Rd}} + \frac{\alpha_y M_y}{M_{b,\,Rd}} + \frac{\alpha_z M_z}{M_{el,\,z}} \leq 1 \tag{3.8}$$

Example 3.2: Combined moments and compression applied to a hollow section

The walls of an excavation are propped using rectangular hollow sections that have a yield stress of 355 N/mm² and that weigh 0.725 kN per meter length see (Figure 3.6).

1. Determine the elastic bending strengths for the prop about the strong and weak axes.
2. Determine the elastic critical buckling forces for each axis of buckling.
3. Determine the amplification of moment factors for each axis of buckling if the applied axial force = 450 kN.
4. Determine the factor of safety (FoS) against buckling if the prop resists a compression force of 450 kN.
5. Determine the FoS if the prop is also subjected to an accidental sideways force of 70 kN applied at midspan.

Note: Calculate the FoS using unfactored loads (see Equation 1.11).

1. The first step is to calculate the second moment of area about the strong and weak axes, which are calculated by taking away I of the centre void from the I for a solid section, i.e.,

$$I_y = \frac{200 \times 300^3}{12} - \frac{180 \times 280^3}{12} = 121 \times 10^6 \, mm^4$$

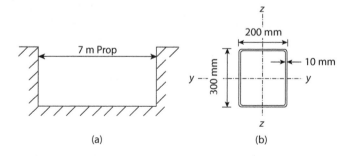

Figure 3.6 Design of props for an excavation. (a) Section through excavation and (b) section through prop.

and

$$I_z = \frac{300 \times 200^3}{12} - \frac{280 \times 180^3}{12} = 64 \times 10^6 \, \text{mm}^4$$

and the elastic moment capacities are

$$M_{\text{el}, y} = \frac{f_y I_y}{D/2} = \frac{355 \times 121 \times 10^6}{300/2} \times 10^{-6} = 286 \, \text{kN.m}$$

and

$$M_{\text{el}, z} = \frac{f_y I_z}{B/2} = \frac{355 \times 64 \times 10^6}{200/2} \times 10^{-6} = 227 \, \text{kN.m}$$

2. If the supports are pinned, the buckling mode is the same as that illustrated in Figure 3.1a and the effective length equals 7000 mm about both axes. From Equation 3.1

$$N_{\text{cr}, y} = \frac{\pi^2 \times 210000 \times 121 \times 10^6}{7000^2} \times 10^{-3} = 5118 \, \text{kN}$$

and

$$N_{\text{cr}, z} = \frac{\pi^2 \times 210000 \times 64 \times 10^6}{7000^2} \times 10^{-3} = 2707 \, \text{kN}$$

3. From Equation 3.5, the amplification factors are

$$\alpha = \frac{1}{1 - \dfrac{N}{N_{\text{cr}}}}$$

$$\alpha_y = \frac{1}{1 - \dfrac{450}{5118}} = 1.10$$

and

$$\alpha_z = \frac{1}{1 - \dfrac{450}{2707}} = 1.20$$

4. The cross-sectional area of the section is

$$\text{Area} = 300 \times 200 - (300 - 20) \times (200 - 20) = 9600 \, \text{mm}^2$$

and from Equation 3.3, the crushing force is

$$N_{\text{pl}, \text{Rd}} = 355 \times 9600 \times 10^{-3} = 3408 \, \text{kN}$$

The strut is free to buckle about either the strong y–y or weak z–z axis. Therefore, buckling will be about the weak axis and from Equation 3.4,

$$N_{b,\,Rd} = \left(\frac{1}{3408} + \frac{1}{2707} \right)^{-1} = 1509 \text{ kN}$$

The applied moment due to the self-weight is

$$M_y = \frac{0.725 \times 7^2}{8} = 4.44 \text{ kN.m}$$

The moment about the z–z axis is zero. Inputting the variables into Equation 3.6

$$\frac{N}{N_{b,\,Rd}} + \frac{\alpha_y M_y}{M_{el,\,y}} + \frac{\alpha_z M_z}{M_{el,\,z}} \leq 1$$

$$\frac{450}{1509} + \frac{1.10 \times 4.44}{286} + \frac{1.20 \times 0}{227} = 0.315$$

The sum of this equation represents the degree of utilisation, which in this case is 31.5%. The FoS is the inverse, i.e.,

$$\text{FoS} = 0.315^{-1} = 3.2$$

5. The midspan moment in a beam of length L subjected to a point load P at midspan is

$$M = \frac{PL}{4}$$

and the moment due to the accidental horizontal force is

$$M_z = \frac{70 \times 7}{4} = 123 \text{ kN.m}$$

Inputting the results into Equation 3.8, we get

$$\frac{450}{1509} + \frac{1.10 \times 4.44}{286} + \frac{1.20 \times 123}{227} = 0.96 \tag{3.9}$$

Therefore, the FoS is

$$\text{FoS} = 0.96^{-1} = 1.04$$

In summary, the FoS has fallen from 3.2 to 1.04, and this illustrates the vulnerability of slender compression members to accidental loading. This 70 kN load could easily have resulted from a careless crane driver accidentally hitting the prop with a skip of concrete on a windy day.

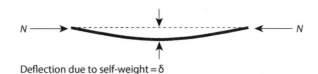

Deflection due to self-weight = δ

Figure 3.7 Illustration of additional (N–δ) moment due to self-weight deflection.

This analysis included no allowance for moments induced by the propping force multiplied by the self-weight deflection. This is known as a N–δ moment and is illustrated in Figure 3.7. In this example, the beam deflects by only 0.9 mm under its own self-weight; therefore, this effect is negligible.

Example 3.3: Beam column with lateral torsional buckling

An H-beam with properties listed below is used to prop an excavation as shown in Figure 3.8.

1. Determine the moment capacity when bending about the strong axis assuming the beam is torsionally restrained at the supports.
2. Determine the elastic critical buckling forces and corresponding amplification of moment's factors about each axis of buckling, if the prop resists a 470 kN axial force.
3. Determine the buckling strength of the prop in pure compression (no moments).
4. Determine the FoS if the prop is also subjected to an accidental sideways force of 70 kN applied at midspan.

Note: Calculate the FoS using unfactored loads in accordance with Equation 1.11.

Section properties for the H-section shown in Figure 3.8b: $W_{pl,y} = 2680$ cm³, $W_{el,y} = 808$ cm³, $I_y = 38{,}750$ cm⁴, $I_z = 12{,}570$ cm⁴, $I_t = 378$ cm⁴, $A = 201$ cm², self-weight = 1.52 kN/m, $f_y = 355$ N/mm², $E = 210{,}000$ N/mm² and $G = 80{,}770$ N/mm²

1. Since the beam column is unrestrained against the weak axis, it will be prone to lateral torsional buckling. From Equation 2.19, the elastic critical buckling moment is

$$M_{cr} = \frac{\pi}{L_{cr}} \sqrt{\frac{EI_z GI_t}{1 - \dfrac{I_z}{I_y}}}$$

$$M_{cr} = \frac{\pi}{7000} \times 10^{-6} \sqrt{\frac{210000 \times 12570 \times 10^4 \times 80770 \times 378 \times 10^4}{1 - \dfrac{12570}{38750}}} = 1550 \text{ kN.m}$$

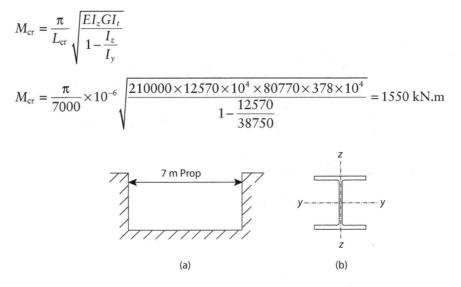

(a)

(b)

Figure 3.8 Prop supporting an excavation that is susceptible to LTB. (a) Section through excavation and (b) section through prop.

From Equation 2.9, the plastic moment capacity is

$$M_{pl,\,Rd} = f_y W_{pl}$$

$$M_{pl,\,Rd} = 2680 \times 10^3 \times 355 \times 10^{-6} = 951 \text{ kN.m}$$

And from Equation 2.21, the bending moment capacity inclusive of lateral torsional buckling is

$$M_{b,\,Rd} = \left(\frac{1}{M_{pl,\,Rd}} + \frac{1}{M_{cr}} \right)^{-1}$$

$$M_{b,\,Rd} = \left(\frac{1}{951} + \frac{1}{1550} \right)^{-1} = 589 \text{ kN.m}$$

2. From Equation 3.1, the elastic critical forces are

$$N_{cr,\,y} = \frac{\pi^2 \times 210000 \times 38750 \times 10^4}{7000^2} \times 10^{-3} = 16390 \text{ kN}$$

and

$$N_{cr,\,z} = \frac{\pi^2 \times 210000 \times 12570 \times 10^4}{7000^2} \times 10^{-3} = 5317 \text{ kN}$$

From Equation 3.5, the amplification factors are

$$\alpha_y = \frac{1}{1 - \dfrac{470}{16390}} = 1.030$$

and

$$\alpha_z = \frac{1}{1 - \dfrac{470}{5317}} = 1.100$$

3. From Equation 3.3, the crushing strength is

$$N_{pl,\,Rd} = 355 \times 20100 \times 10^{-3} = 7136 \text{ kN}$$

and the buckling strength, using the weak axis elastic critical buckling force and Equation 3.4, is

$$N_{b,\,Rd} = \left(\frac{1}{N_{pl,\,Rd}} + \frac{1}{N_{cr}} \right)^{-1}$$

$$N_{b,\,Rd} = \left(\frac{1}{7136} + \frac{1}{5317} \right)^{-1} = 3047 \text{ kN}$$

4. And the moment due to the accidental horizontal force applied at midspan is

$$M_z = \frac{PL}{4} = \frac{70 \times 7}{4} = 123 \text{ kN.m}$$

The self-weight induced midspan moment is

$$M_y = \frac{1.52 \times 7^2}{8} = 9.3 \text{ kN.m}$$

From Equation 2.10, the weak axis elastic moment capacity is

$$M_{el,\,Rd} = f_y W_{el}$$

$$M_{el,\,z} = 808 \times 10^3 \times 355 \times 10^{-6} = 287 \text{ kN.m}$$

Finally, the beam column interaction equation (Equation 3.8) is solved:

$$\frac{N}{N_{b,\,Rd}} + \frac{\alpha_y M_y}{M_{b,\,Rd}} + \frac{\alpha_z M_z}{M_{el,\,z}} = 1.0$$

$$\frac{470}{3047} + \frac{1.030 \times 9.3}{589} + \frac{1.100 \times 123}{287} = 0.64$$

The FoS is

$$\text{FoS} = 0.64^{-1} = 1.56$$

To summarise, the prop can resist a 70 kN accidental force with a reasonable FoS of 1.56. These calculations did not include load factors, although 1.56 is high enough to accommodate load factors and still satisfy the limit state design requirements.

3.3 WEB BUCKLING

Beam webs are vulnerable to buckling under concentrated forces and the strength can be estimated by using strut buckling theory. To do this, it is necessary to estimate the effective length, which is determined in exactly the same manner as for any ordinary strut (see Figure 3.1). For the common condition where the top and bottom flanges are not free to rotate or move sideways, then $L_{cr} = 0.5 \, D$ (Figure 3.9a). This end condition produces an elastic critical buckling strength four times that of the web shown in Figure 3.9b and 16 times that of Figure 3.9c because of the inverse squared relationship between buckling force and length (Equation 3.1). For this reason, it is very important to establish the correct effective

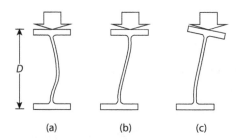

Figure 3.9 Different modes of web buckling. (a) $L_{cr} = 0.5\,D$, (b) $L_{cr} = D$ and (c) $L_{cr} = 2\,D$.

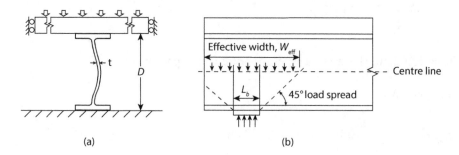

Figure 3.10 Principle of load spreading illustrated using a bridge bearing. (a) Cross section showing buckling and (b) side elevation showing load spreading away from the bearing.

length during calculations, as well as to ensure that appropriate restraints to buckling are provided during construction.

Buckling occurs when the stress at the centre of the buckle exceeds a critical value. Therefore, the next objective is to determine the stress at the centre of the buckle, which occurs at the centre line of the beam web for the case shown in Figure 3.10a.

The stress from a concentrated force is spread at an angle of approximately 45 degrees away from the edges of a stiff bearing (L_b), as shown in Figure 3.10b. For simplicity, it is assumed that the concentrated force is spread evenly across the effective width (W_{eff}), which is calculated using this load dispersion approximation as shown in Figure 3.10b. The concentrated force is assumed to be resisted only by the beam web within the effective width region, with the remainder of the beam ignored. The second moment of area of the web within the effective width is

$$I_{eff} = \frac{w_{eff} \times t^3}{12} \tag{3.10}$$

Inputting this into Equation 3.1, the elastic critical buckling force is

$$N_{cr} = \frac{\pi^2 E I_{eff}}{L_{cr}^2} \tag{3.11}$$

Adapting Equation 3.3, the 'crushing' strength of the web is

$$N_{pl,\,Rd} = f_y w_{eff} t \tag{3.12}$$

The design buckling strength is calculated using the Gordon–Rankine approximation (Equation 3.4), i.e.,

$$N_{b,\,Rd} = \left(\frac{1}{N_{pl,\,Rd}} + \frac{1}{N_{cr}} \right)^{-1}$$

Example 3.4: Web buckling due to a concentrated load

I-section beams support a reinforced concrete slab, as shown in Figure 3.11. Determine the buckling strength when subjected to the concentrated force of the support reaction at the ends of the beam. The top and bottom flanges are restrained against sideways movement; the yield stress is 355 N/mm² and $E = 210,000$ N/mm².

The beam flanges prevent rotation of the web at the supports and sideways movement of the slab is prevented (Figure 3.12a); therefore, the effective length of the web is

$$L_{cr} = 0.5\ D = 250\ \text{mm}$$

The load spreading is illustrated in Figure 3.12b and

$$W_{eff} = 160 + 150 + 250 = 560\ \text{mm}$$

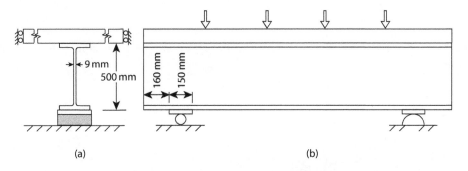

Figure 3.11 Web buckling. (a) Cross section and (b) side elevation.

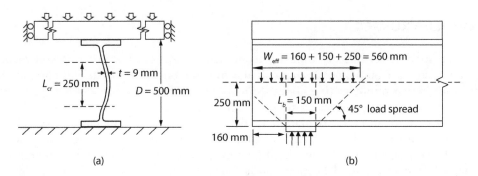

Figure 3.12 Web buckling at a bearing. (a) Cross section showing effective length and (b) side elevation showing calculation of effective width.

The crushing force from Equation 3.12 is

$$N_{pl,\,Rd} = f_y w_{eff} t$$

$$N_{pl,\,Rd} = 355 \times 560 \times 9 \times 10^{-3} = 1789 \text{ kN}$$

The second moment of area of the effective width of the web from Equation 3.10 is

$$I_{eff} = \frac{w_{eff} \times t^3}{12}$$

$$I_{eff} = \frac{560 \times 9^3}{12} = 34020 \text{ mm}^4$$

And the elastic critical buckling force from Equation 3.11 is

$$N_{cr} = \frac{\pi^2 E I_{eff}}{L_{cr}^2}$$

$$N_{cr} = \frac{\pi^2 \times 210000 \times 34020}{250^2} \times 10^{-3} = 1128 \text{ kN}$$

Finally, from Equation 3.4 the web can (probably) resist the following reaction without buckling:

$$N_{b,\,Rd} = \left(\frac{1}{N_{pl,\,Rd}} + \frac{1}{N_{cr}} \right)^{-1}$$

$$N_{b,\,Rd} = \left(\frac{1}{1789} + \frac{1}{1128} \right)^{-1} = 692 \text{ kN}$$

3.4 SIMPLE TRUSSES

Many long-span structures are built up from struts and ties. The compression members must be checked for buckling. During calculations, the effective length will correspond to the distance between intersecting nodes *if the nodes are braced against sideways moment*, as is the case for the truss sketched in Figure 3.13.

Computers can be used to calculate truss deflections, although the parallel axis theorem can be used to estimate the effective second moment of area of a truss, for use in approximate deflection calculations. The simplification ignores shear deflections; therefore, the solution will underestimate deflection and is thus only suitable as a 'back of the envelope' check. Shear deflections are caused by the stretching and squashing of the diagonal web members due to tension and compression forces.

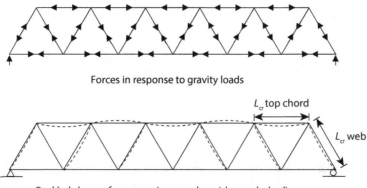

Forces in response to gravity loads

L_{cr} top chord

L_{cr} web

Buckled shape of compression members (shown dashed)

Figure 3.13 Lattice girder in which nodes between members are braced against sideways movement.

Example 3.5: Lattice girder bridge

The lattice girder sketched in Figure 3.14 is constructed out of steel with a yield stress of 355 N/mm² and Young's modulus of 210,000 N/mm². All the truss members are constructed out of 200 mm wide square hollow sections, with a cross-sectional area of 87 cm² and I of 4860 cm⁴.

1. The dead load shown in Figure 3.14 includes the self-weight. Determine the ULS design load, midspan moment and maximum compression force in the top chord of the truss.
2. If the top chord is restrained against sideways movement at node points (i.e. the intersections between members), determine the effective length of the top chord of the truss.
3. Determine if the top chord is capable of resisting the applied ULS compression force without buckling.
4. Calculate the SLS design load (inclusive of dead weight) and estimate the midspan deflection. If the maximum deflection limit is span/200, is the truss satisfactory?

1. From Equation 1.3, the design load is

$$w = 1.35 \times 12 + 1.5 \times 10 = 31.2 \text{ kN/m}$$

and the midspan moment is

$$M = \frac{wL^2}{8} = \frac{31.2 \times 48^2}{8} = 8985 \text{ kN.m}$$

This is resisted by a couple between the top and bottom chords, in a manner illustrated in Figure 3.15.

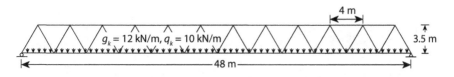

$g_k = 12$ kN/m, $q_k = 10$ kN/m

4 m

3.5 m

48 m

Figure 3.14 Lattice girder design example.

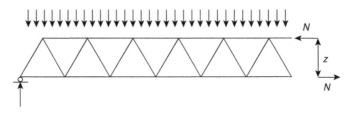

Figure 3.15 Balance of forces at midspan.

Taking moments about the bottom chord (see Figure 3.15)

$$M = N \times z \tag{3.13}$$

where
  N is the force in the top chord.
  Z is the distance between the centre of the top and bottom chords.

Rearranging Equation 3.13

$$N = \frac{M}{z} \tag{3.14}$$

In this case, the midspan moment produces a force in the chord of

$$N = \frac{8985}{3.5} = 2567 \text{ kN}$$

2. The truss will buckle in a manner similar to that illustrated in Figure 3.13 if the nodes are prevented from buckling sideways. If this is the case, then the effective length is the distance between web members, i.e., $L_{cr} = 4$ m.

3. From Equation 3.3, the crushing strength of the chord members is

$$N_{pl, Rd} = f_y A = 355 \times 87 \times 10^2 \times 10^{-3} = 3088 \text{ kN}$$

From Equation 3.1

$$N_{cr} = \frac{\pi^2 \times 210000 \times 4860 \times 10^4}{4000^2} \times 10^{-3} = 6296 \text{ kN}$$

From Equation 3.4, the compression strength is

$$N_{b, Rd} = \left(\frac{1}{3088} + \frac{1}{6296} \right)^{-1} = 2071 \text{ kN}$$

Since the applied force (2567 kN) is greater than the buckling strength (2071 kN), the top chord is *not strong enough*.

These calculations ignore the N–δ bending moments induced by the self-weight deflection of the top chord (see Figure 3.7), although in this case the top chord sags by less than 1/2 mm between nodes, so the loss of strength is insignificant.

4. Using the parallel axis theorem, the effective second moment of area of the truss is

$$I = \sum \left(I_{chord} + \text{Area}_{chord} \cdot r^2 \right) \tag{3.15}$$

where r is the distance from the truss centroid to the chord centroid; therefore,

$$I = 2 \times \left(4860 \times 10^4 + 87 \times 10^2 \times 1750^2 \right) = 5.34 \times 10^{10} \, \text{mm}^4$$

The SLS load is

$$w = g_k + q_k = 22 \text{ kN/m or } 22 \text{ N/mm}$$

The midspan deflection (δ) in a beam supporting a UDL is

$$\delta = \frac{5wL^4}{384EI}$$

$$\delta = \frac{5 \times 22 \times 48000^4}{384 \times 210000 \times 5.34 \times 10^{10}} = 140 \text{ mm}$$

This answer was rounded up to the nearest 10 mm to reflect the approximate nature of the calculation. The max allowable deflection is $L/200 = 240$ mm; therefore, the truss is stiff enough, even though it has insufficient strength. In practice, the actual deflection will be slightly higher than 140 mm, because this simplified approach ignores shear deflections. These are caused by the stretching and squashing of the diagonal web members.

Example 3.6: Lattice girder roof

A truss supports a dead load inclusive of self-weight of 10 kN/m (unfactored) in addition to an imposed load of 60 kN (unfactored) (see Figure 3.16). The yield stress is 275 N/mm² and Young's modulus is 210,000 N/mm².

1. Determine the ULS compression force in the top chord of the truss at midspan.
2. The truss members are fabricated from 300 mm wide square hollow sections with a wall thickness of 10 mm. Determine the area, second moment of area and elastic moment capacity of the top chord.
3. Determine the compression force required to cause buckling of the top chord if it is restrained against sideways movement.
4. The uniformly distributed dead load is supported directly by the top chord. It therefore induces bending moments in addition to the axial compression. Determine the maximum moment that can be resisted in addition to the applied compression force.
5. Determine the maximum midspan deflection under serviceability limit state loading.

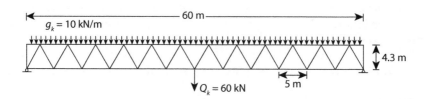

Figure 3.16 Roof truss.

1. The midspan moment due to the fully factored dead is

$$M = \frac{wL^2}{8} + \frac{PL}{4}$$

$$M = \frac{1.35 \times 10 \times 60^2}{8} + \frac{1.5 \times 60 \times 60}{4} = 7425 \text{ kN.m}$$

From Equation 3.13, the compression force in the top chord at midspan is

$$N = \frac{7425}{4.3} = 1726.7 \text{ kN}$$

2. The cross-sectional area of the top chord is

$$A = 300 \times 300 - 280 \times 280 = 11600 \text{ mm}^2$$

and the second moment of area is

$$I = \frac{300 \times 300^3}{12} - \frac{280 \times 280^3}{12} = 162.79 \times 10^6 \text{ mm}^4$$

and the elastic moment capacity from Equation 2.13 is

$$M_{el, Rd} = \frac{275 \times 162.79 \times 10^6}{150} \times 10^{-6} = 298.4 \text{ kN.m}$$

3. From Equation 3.3, the crushing resistance is

$$N_{pl, Rd} = f_y A = 275 \times 11600 \times 10^{-3} = 3190 \text{ kN}$$

The effective length is the distance between the nodes with the web members (5 m), since the roof provides bracing against sideways movements. From Equation 3.1, the elastic critical buckling force is

$$N_{cr} = \frac{\pi^2 \times 210000 \times 162.79 \times 10^6}{5000^2} \times 10^{-3} = 13496 \text{ kN}$$

and from Equation 3.4, the buckling strength is

$$N_{b,\text{Rd}} = \left(\frac{1}{3190} + \frac{1}{13496} \right)^{-1} = 2580 \text{ kN}$$

This is greater than the applied force of 1726.7 kN; therefore, this is encouraging, although no allowance has been made for the moment in the chord due to the UDL supported.

4. The amplification of moment's factor from Equation 3.5 is

$$\alpha_z = \frac{1}{1 - \dfrac{1726.7}{13496}} = 1.15$$

and the main interaction equation (Equation 3.6) is

$$\frac{N}{N_{b,\text{Rd}}} + \frac{\alpha_y M_y}{M_{\text{el},y}} + \frac{\alpha_z M_z}{M_{\text{el},z}} \leq 1$$

$$\frac{1726.7}{2580} + \frac{1.15 \times M_y}{298.4} + \frac{1.15 \times 0}{298.4} \leq 1.0$$

$M_y \leq 85.8$ kNm.

A quick calculation assuming conservatively that the moment in the top chord was $wL^2/8$ (where $L = 5$ m) shows that $M_y < 49$ kN.m. Therefore, the top chord should be strong enough.

A slight reduction in strength will occur due to the moment induced by the self-weight deflection, as illustrated in Figure 3.7. In this example, the top chord sags by only 2 mm between nodes and the resulting (N–δ) moment is not significant.

5. The second moment of area of the truss from Equation 3.15 is

$$I_{\text{truss}} = 2 \times \left(162.79 \times 10^6 + 11600 \times 2150^2 \right) = 1.076 \times 10^{11} \text{ mm}^4$$

The SLS loads are a UDL of 10 N/mm and a point load of 60×10^3 N, and the corresponding midspan deflection is

$$\Delta = \frac{5wL^4}{384EI} + \frac{PL^3}{48EI}$$

$$\Delta = \frac{5 \times 10 \times 60000^4}{384 \times 210000 \times 1.076 \times 10^{11}} + \frac{60 \times 10^3 \times 60000^3}{48 \times 210000 \times 1.076 \times 10^{11}} = 90 \text{ mm}$$

This is rounded up to the nearest 10 mm because of the approximate nature of the calculation. This deflection is unconservative, because it ignores the deflection that occurs due to the stretching and squashing of the web members, known as *shear deflection*.

3.5 BUCKLING OF SLENDER TRUSSES

Laced props comprises an array of different members (Figure 3.17a). These compound members have two modes of buckling: local buckling as sketched in Figure 3.17b and global buckling (Figure 3.17c). The actual buckling mode is a combination of both modes as illustrated in Figure 3.17d. Whilst a detailed computer-based analysis of these structures is desirable, a simple approximation of the buckling strength is possible with hand calculations. The local buckling force, N_{local}, shown in Figure 3.17b, replaces the crushing force in a conventional strut. The chord members will buckle at a load lower than the yield (crushing) load. Therefore, N_{local} (for the truss as a whole) is the sum of the buckling strength of the individual chord members, calculated using the effective length shown in Figure 3.17b.

The elastic critical buckling force for the global buckling mode shown in Figure 3.17c is determined using the second moment of area of the whole lattice member. This can be determined using the parallel axis theorem (Equation 3.15). The elastic critical buckling force can be calculated using Equation 3.1. The combined failure load, $N_{b,Rd}$, (Figure 3.17d) is estimated using the combined local and global buckling modes, i.e.,

$$\frac{1}{N_{b,Rd}} = \frac{1}{N_{local}} + \frac{1}{N_{cr,global}}$$

Rearranging

$$N_{b,Rd} = \left(\frac{1}{N_{local}} + \frac{1}{N_{cr,global}} \right)^{-1} \tag{3.16}$$

Deformation of the diagonal web members (known as *shear deflection*) will increase lateral deflections, reducing N_{cr}. Although Equation 3.1 is sufficiently accurate for what is already an approximate method, greater accuracy can be achieved by using equations published

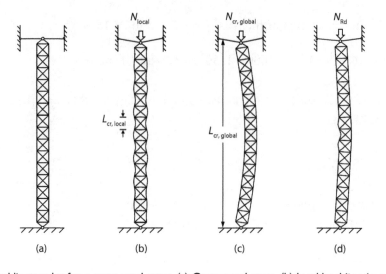

Figure 3.17 Buckling modes for a compound strut. (a) Compound strut, (b) local buckling (restrained against global buckling), (c) global (elastic critical) buckling and (d) real behaviour, modes (b) and (c) combined.

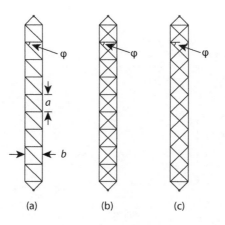

Figure 3.18 Types of compound strut.

in Timeshenko and Gere (1961). These include the effect of elastic deformation of the web members. For the compound strut shown in Figure 3.18a, the elastic critical buckling force is

$$N_{cr, global} = \frac{\pi^2 EI}{L_{cr}^2} \frac{1}{1 + \dfrac{\pi^2 EI}{L_{cr}^2} \left(\dfrac{1}{A_d E \sin \varphi \cos^2 \varphi} + \dfrac{b}{a A_c E} \right)} \tag{3.17}$$

A_d is the area of two diagonal web members, one on each side of the compound strut (Figure 3.18a).
A_c is the combined area of the main chord members.
For the member types shown in Figure 3.18b and c, the elastic critical buckling force is

$$N_{cr, global} = \frac{\pi^2 EI}{L_{cr}^2} \frac{1}{1 + \dfrac{\pi^2 EI}{L_{cr}^2} \left(\dfrac{1}{A_d E \sin \varphi \cos^2 \varphi} \right)} \tag{3.18}$$

A_d is the cross-sectional area of four diagonal members, two on each side of the compound strut.

Example 3.7: Buckling of a lattice girder in compression

A 19 m long prop is sketched in Figure 3.19a. The yield stress is 355 N/mm² and Young's modulus is 210,000 N/mm².

1. Determine the local buckling force (Figure 3.17b).
2. Determine the elastic critical buckling force of the prop using the approximate method, i.e., not the Timoshenko method (Figure 3.17c).
3. Determine the design compression strength.
4. The diagonal web members are 25 mm wide square hollow section members with a wall thickness of 3 mm. Recalculate the compression strength using Timoshenko's equation (Equation 3.18).

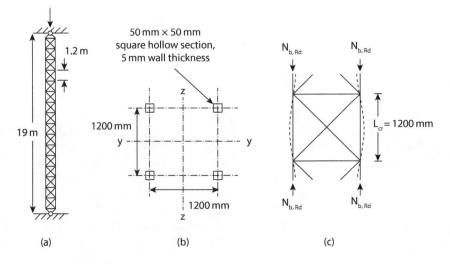

Figure 3.19 Lattice girder in compression. (a) Elevation, (b) cross section and (c) local buckling.

1. The cross-sectional area of an individual chord member is

$$\text{Area} = 50^2 - (50 - 2 \times 5)^2 = 900 \text{ mm}^2$$

From Equation 3.3, the crushing force is

$$N_{\text{pl, Rd}} = 355 \times 900 \times 10^{-3} = 319.5 \text{ kN}$$

and the second moment of area is

$$I = \frac{50 \times 50^3}{12} - \frac{40 \times 40^3}{12} = 307.5 \times 10^3 \text{ mm}^4$$

For local buckling, the effective length, L_{cr} = 1.0 L = 1200 mm (see Figure 3.19c) and from Equation 3.1

$$N_{\text{cr, local}} = \frac{\pi^2 \times 210000 \times 307.5 \times 10^3}{1200^2} \times 10^{-3} = 443 \text{ kN}$$

and the design compression strength of a single chord member from Equation 3.4 is

$$N_{b,\text{Rd}} = \left(\frac{1}{319.5} + \frac{1}{443}\right)^{-1} = 186 \text{ kN}$$

The local buckling mode (Figure 3.17b) requires all four-leg members to buckle; therefore,

$$N_{\text{local}} = 4 \times N_{b,\text{Rd}} \tag{3.19}$$

$$N_{\text{local}} = 4 \times 186 = 744 \text{ kN}$$

2. From Equation 3.15, the second moment of area of the lattice girder is

$$I = 4 \times \left(307.5 \times 10^3 + 900 \times 600^2\right) = 1.3 \times 10^9 \, \text{mm}^4$$

The effective length of the global buckling model (Figure 3.17c) is 19 m. From Equation 3.1,

$$N_{\text{cr, global}} = \frac{\pi^2 \times 210000 \times 1.3 \times 10^9}{19000^2} \times 10^{-3} = 7464 \, \text{kN}$$

3. The design strength from Equation 3.16 is

$$N_{\text{Rd}} = \left(\frac{1}{N_{\text{local}}} + \frac{1}{N_{\text{cr, global}}}\right)^{-1} = \left(\frac{1}{744} + \frac{1}{7464}\right)^{-1} = 677 \, \text{kN}$$

4. The cross-sectional area of a web member is

$$\text{Area} = 25^2 - (25 - 2 \times 3)^2 = 264 \, \text{mm}^2$$

A_d is the combined area of the diagonal web members = 4 × 264 = 1056 mm².

ϕ is the internal angle of the web members, which in this case is 45° (see Figure 3.18b).

It has already been established that the basic value for the elastic critical buckling force is 7464 kN. From Equation 3.18,

$$N_{\text{cr, global}} = 7464 \times 10^3 \times \frac{1}{1 + 7464 \times 10^3 \times \left(\dfrac{1}{1056 \times 210000 \times \sin 45 \cos^2 45}\right)} \times 10^{-3}$$

$$= 6815 \, \text{kN}$$

[Note that an elastic finite element analysis (FEA) solution to this problem predicted an elastic critical buckling force of 6940 kN.]

The design strength from Equation 3.16 is

$$N_{\text{Rd}} = \left(\frac{1}{744} + \frac{1}{6815}\right)^{-1} = 671 \, \text{kN}$$

This shows that shear deflections only reduced the design strength from 677 kN to 671 kN. An elastic–plastic FEA solution showed a failure load of 1030 kN, which is significantly higher than 671 kN. However, the FEA solution is a mathematic prediction, ignoring residual stresses and other imperfections. It will therefore overestimate strength.

3.6 BUCKLING OF SLENDER TRUSSES SUBJECTED TO COMPRESSION AND BENDING

It is possible to estimate the strength of lattice girders when subjected to combined compression and bending by extending the method used for conventional beam columns *except* that the elastic moment capacity is replaced by the moment capacity limited by the buckling of the chord members. This will occur at a stress below the yield stress (see Figure 3.20).

The reduced moment capacity is equal to the buckling force in the chords multiplied by the lever arm distance between the chords. The combined bending and compression check is

$$\frac{N}{N_{Rd}} + \frac{\alpha_y M_y}{M_{Rd,\,y}} + \frac{\alpha_z M_z}{M_{Rd,\,z}} \le 1 \tag{3.20}$$

where $M_{Rd,\,y}$ and $M_{Rd,\,z}$ are the leg moment capacities about the strong and weak axes, respectively, calculated using the product of the buckling force of the leg members and the lever arm. This method for estimating the bending strength is accurate, providing the diagonal web members are strong enough to carry the shear forces. It would not work for Vierendeel girders.

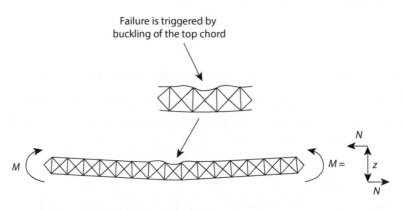

Figure 3.20 Moment capacity calculation method.

Example 3.8: Lattice girder subjected to compression and bending

Determine the FoS, if the lattice girder considered in Example 3.7 supports a propping force of 200 kN combined with a bending moment of 50 kN.m about the *y–y* axis and 25 kN.m about the *z–z* axis, all unfactored. The bending axes are shown in Figure 3.19.

It was shown in Example 3.7 that the chords of the truss each have a compression strength of 186 kN. Using the method illustrated in Figure 3.20, the moment capacity from Equation 3.13 is

$$M = N \times z$$

In this case, the chord members were shown to buckle at 186 kN; therefore, the bending strength is

$$M_{Rd} = 2 \times 186 \times 1.2 = 446 \text{ kN.m}$$

The objective is to determine the FoS, which is described in Section 1.3. This is a measure of the amount of overload the structure can take, and it is determined using unfactored loads. From Equation 3.5, the amplification of moments factor is

$$\alpha = \frac{1}{1 - \dfrac{N}{N_{cr}}}$$

$$\alpha = \frac{1}{1 - \dfrac{200}{7464}} = 1.03$$

From Equation 3.20

$$\frac{N}{N_{Rd}} + \frac{\alpha_y M_y}{M_{Rd,\,y}} + \frac{\alpha_z M_z}{M_{Rd,\,z}} \leq 1$$

$$\frac{186}{677} + \frac{1.03 \times 50}{446} + \frac{1.03 \times 25}{446} = 0.448$$

The FoS is the inverse, i.e.,

$$FoS = 0.448^{-1} = 2.23$$

Example 3.9: Lattice girder propping an excavation

A construction is propped using a series of struts, one of which is sketched in Figure 3.21. The steel has a yield stress of 355 N/mm² and Young's modulus of 210,000 N/mm².

1. Determine the compression strength of the prop in the absence of moments.
2. Determine the bending moment capacity of the prop.
3. Determine the FoS (against unfactored loads) if the prop resists a compression force of 1000 kN in addition to a self-weight of 2.4 kN/m and an accidental sideways force of 50 kN applied at midspan

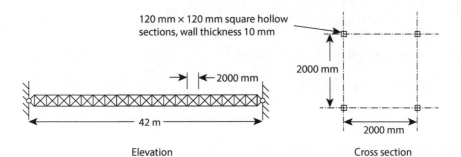

Figure 3.21 Prop for an excavation.

1. The cross-sectional area of a chords is

$$\text{Area} = 120^2 - 100^2 = 4400 \text{ mm}^2$$

and the second moment of area is

$$I = \frac{120 \times 120^3}{12} - \frac{100 \times 100^3}{12} = 8.95 \times 10^6 \text{ mm}^4$$

From Equation 3.3, the crushing force is

$$N_{\text{pl, Rd}} = f_y A = 355 \times 4400 \times 10^{-3} = 1562 \text{ kN}$$

For local buckling, the effective length is L_{cr} = 1.0 L = 2000 mm, and from Equation 3.1

$$N_{\text{cr, local}} = \frac{\pi^2 \times 210000 \times 8.95 \times 10^6}{2000^2} \times 10^{-3} = 4637 \text{ kN}$$

From Equation 3.4, the capacity per leg of the strut is

$$N_{b,\text{ Rd}} = \left(\frac{1}{1562} + \frac{1}{4637} \right)^{-1} = 1168 \text{ kN}$$

From Equation 3.19, the compression strength for the buckling mode shown in Figure 3.17b is

$$N_{\text{local}} = 4 \times 1168 = 4672 \text{ kN}$$

From Equation 3.15, the second moment of area of the prop is

$$I = 4 \times \left(8.95 \times 10^6 + 4400 \times 1000^2 \right) = 1.764 \times 10^{10} \text{ mm}^4$$

The effective length of the global buckling mode (Figure 3.17c) is 42 m. From Equation 3.1

$$N_{\text{cr}} = \frac{\pi^2 \times 210000 \times 1.764 \times 10^{10}}{42000^2} \times 10^{-3} = 20726 \text{ kN}$$

The compression strength (with no moments) from Equation 3.16 is

$$N_{\text{Rd}} = \left(\frac{1}{4672} + \frac{1}{20726} \right)^{-1} = 3812 \text{ kN}$$

2. From Equation 3.13, the moment capacity is

$$M_{\text{Rd}} = 2 \times 1168 \times 2.0 = 4672 \text{ kN.m}$$

3. The midspan moment (in the vertical plane) due to self-weight is

$$M_y = \frac{2.4 \times 42^2}{8} = 529 \text{ kN.m}$$

The midspan moment (in the horizontal plane) due to the accidental force is

$$M_z = \frac{50 \times 42}{4} = 525 \text{ kN.m}$$

From Equation 3.5, the amplification factor is

$$\alpha = \frac{1}{1 - \dfrac{1000}{20721}} = 1.05$$

Finally, the main interaction equation (Equation 3.20) is solved:

$$\frac{N}{N_{Rd}} + \frac{\alpha_y M_y}{M_{Rd,\,y}} + \frac{\alpha_z M_z}{M_{Rd,\,z}} \leq 1$$

$$\frac{1000}{3812} + \frac{1.05 \times 529}{4672} + \frac{1.05 \times 525}{4672} = 0.50$$

and the FoS against unfactored loads is

$$FoS = 0.50^{-1} = 2.0$$

This should probably be sufficient; however, the calculations ignore the additional moments induced in the chord member if the accidental force is positioned between node points. This will reduce the FoS somewhat, and this effect is investigated in the next example.

Example 3.10: Analysis of impacts on a lattice girder propping a construction

The lattice prop sketched in Figure 3.22 is pinned top and bottom and has a yield stress of 355 N/mm².

1. Determine the buckling force in the absence of bending moments (Figure 3.17d).
2. Estimate the buckling force if the strut is subjected to an accidental force of 70 kN applied at Point A (midheight), shown in Figure 3.22 (assume this is the midspan position).
3. Recalculate the buckling force if the accidental force is applied at Point B.

1. The area of a single chord is

$$\text{Area} = 200^2 - 176^2 = 9024 \text{ mm}^2$$

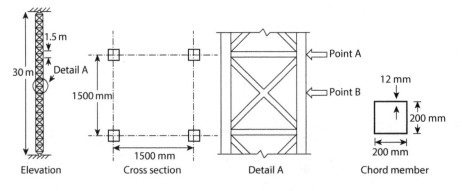

Figure 3.22 Lattice girder prop.

From Equation 3.3

$$N_{pl,\,Rd} = 355 \times 9024 \times 10^{-3} = 3203.5 \text{ kN}$$

The second moment of area of the chords is

$$I = \frac{200 \times 200^3}{12} - \frac{176 \times 176^3}{12} = 53.4 \times 10^6 \text{ mm}^4$$

The effective length of the chord members, L_{cr} = 1.0 L = 1500 mm; therefore, from Equation 3.1

$$N_{cr,\,local} = \frac{\pi^2 \times 210000 \times 53.4 \times 10^6}{1500^2} \times 10^{-3} = 49190 \text{ kN}$$

From Equation 3.4, the basic compression strength of a chord member is

$$N_{b,\,Rd} = \left(\frac{1}{3203.5} + \frac{1}{49190.1} \right)^{-1} = 3008 \text{ kN}$$

From Equation 3.19

$$N_{local} = 4 \times 3008 = 12032 \text{ kN}$$

From Equation 3.15, the second moment of area of the compound strut is

$$I = 4 \times \left(53.4 \times 10^6 + 9024 \times 750^2 \right) = 2.05 \times 10^{10} \text{ mm}^4$$

The effective length, L_{cr} = 1.0 L = 30,000 mm and the elastic critical buckling force is

$$N_{cr,\,global} = \frac{\pi^2 \times 210000 \times 2.05 \times 10^{10}}{30000^2} \times 10^{-3} = 47210 \text{ kN}$$

From Equation 3.16, the basic compression strength of the prop is

$$N_{Rd} = \left(\frac{1}{12032} + \frac{1}{47210} \right)^{-1} = 9588 \text{ kN}$$

2. From Equation 3.13, the bending strength is

$$M_{Rd} = 2 \times 3008 \times 1.5 = 9024 \text{ kN.m}$$

The applied moment due to the 70 kN force is

$$M = \frac{70 \times 30}{4} = 525 \text{ kN.m}$$

Ignoring the amplification of moments due to axial load, i.e., let $\alpha = 1.0$, from Equation 3.20

$$\frac{N}{9588} + \frac{1.0 \times 525}{9024} + 0 = 1$$

which solves to $N = 9030$ kN. Therefore, this moderately sized accidental force has only reduced the load capacity from 9588 kN to 9030 kN; therefore, this appears to be a robust design.

3. Moving the accidental force to Point B will introduce bending moments into one of the chord members. This moment is conservatively equal to

$$M_{Ed} = \frac{PL}{4} = \frac{70 \times 1.5}{4} = 26.25 \text{ kN.m}$$

The next step is to calculate the reduced buckling strength of the chord member, but first the elastic moment capacity of the chord must be calculated from Equation 2.13

$$M_{el, Rd} = \frac{355 \times 53.4 \times 10^6}{100} \times 10^{-6} = 189.6 \text{ kN.m}$$

Using the interaction equation for a beam column (Equation 3.7) and ignoring the amplification of moments and the $N-\delta$ moment due to the deflection in the chord member due to the accidental force

$$\frac{N}{N_{b, Rd}} + \frac{M_y}{M_{el, y}} + \frac{M_z}{M_{el, z}} \leq 1$$

$$\frac{N}{3008} + \frac{1.0 \times 26.25}{189.6} + \frac{1.0 \times 0}{189.6} \leq 1.0$$

$$N = 2592 \text{ kN}$$

From Equation 3.19, the local buckling force is

$$N_{local} = 4 \times 2592 = 10368 \text{ kN}$$

The design strength from Equation 3.16 is

$$N_{Rd} = \left(\frac{1}{10368} + \frac{1}{47210} \right)^{-1} = 8501 \text{ kN}$$

From Equation 3.13, the bending strength of the prop is

$$M_{Rd} = 2 \times 2592 \times 1.5 = 7776 \text{ kN.m}$$

Ignoring the amplification of moments due to axial load

$$\frac{N}{8501} + \frac{1.0 \times 525}{7776} + \frac{1.0 \times 0}{7776} = 1$$

$$N = 7927 \text{ kN}$$

Therefore, the change in the position of the accidental force from Point A to Point B has reduced the load capacity by 12%. This truss has relatively stocky chord members and is therefore less sensitive to accidental loading than a more slender design. The calculations included the following simplifications:

a. The nodes between members were assumed to be pinned, although in practice they will be welded and therefore continuous.
b. The amplification of moment's factors were set at 1.0 throughout.
c. The $N-\delta$ moment due to the deflection in the chord resulting from the accidental force was ignored. In this case, the chord will deflect by only 0.4 mm (between nodes) under the accidental force; therefore, this effect is minimal.

Problems

Solutions to these problems are provided at https://www.crcpress.com/9781498741217

P.3.1. The 10 m long vertical strut whose cross section is sketched in Figure 3.23 resists a working load of 200 kN in axial compression. The yield stress is 355 N/mm² and $E = 210,000$ N/mm². Calculate the axial load capacity and FoS against unfactored loads if the strut has the following support conditions:
a. It is effectively pinned at both ends.
b. It is fixed against rotation at both ends.
c. It is fixed against rotation at both ends and restrained against buckling about the weak axis.
Ans. (a) 203.8 kN, 1.02, (b) 689.6 kN, 3.45 and (c) 2985 kN, 14.9.

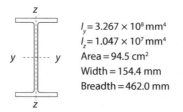

$I_y = 3.267 \times 10^8 \text{ mm}^4$
$I_z = 1.047 \times 10^7 \text{ mm}^4$
Area = 94.5 cm²
Width = 154.4 mm
Breadth = 462.0 mm

Figure 3.23 Section properties of an I-beam.

P.3.2. A bridge comprises two I-section beams supporting a reinforced concrete slab (see Figure 3.24). The slab is loaded by a uniformly distributed load of 6 kN/m².

   a. Determine if the elastic moment capacity of the beam sections exceeds the applied ULS design moment.

   b. Determine the midspan deflection under SLS loads.

   c. Determine if the shear capacity of the beams exceeds the applied shear force.

   d. The bridge is restrained against sideways movement under load. Determine if the buckling capacity of the beam webs is adequate.

I-section beam properties
Self-weight = 0.8 kN/m, depth = 500 mm, web thickness = 9 mm, major axis elastic section modulus = 1800 cm³, major axis second moment of area = 47,500 cm⁴, yield stress = 355 N/mm², Young's modulus = 210,000 N/mm² and reinforced concrete density = 25 kN/m³.

Ans. (a) yes, 639 kN.m > 491.6 kN.m, (b) 36.3 mm, (c) yes, 922.3 kN > 196.5 kN and (d) yes, 556 kN > 196.5 kN.

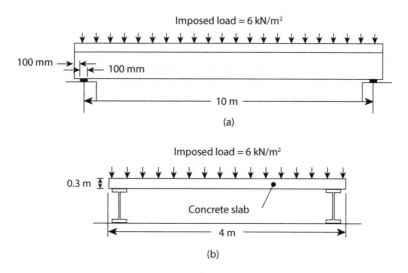

Figure 3.24 Simple bridge. (a) Side elevation and (b) cross section through bridge.

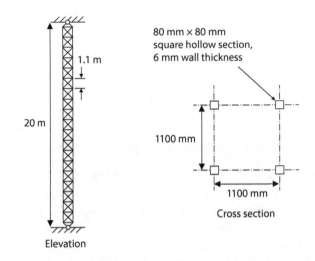

80 mm × 80 mm
square hollow section,
6 mm wall thickness

1.1 m

20 m

1100 mm

1100 mm

Cross section

Elevation

Figure 3.25 Temporary prop.

P.3.3. A contractor intends to use the prop sketched in Figure 3.25 to support a stadium roof during construction. The yield stress is 355 N/mm² and Young's modulus = 210,000 N/mm².

a. Determine the local buckling force for the prop.
b. Determine the second moment of area and the elastic critical buckling force for the prop.
c. Determine the buckling force and estimate the FoS against buckling if the prop supports a compression force of 600 kN.
d. Determine the FoS if the prop is also subjected to an accidental sideways force of 80 kN applied at midheight (at a node).

Ans. (a) 2056 kN, (b) 2.16×10⁹ mm⁴, 11,192 kN, (c) 1737 kN, 2.9 and (d) 1.39.

REFERENCE

Timoshenko, S. P. and Gere, M., 1961. *Theory of Elastic Stability*. 2nd Edition. New York: McGraw-Hill.

Chapter 4

Buckling of arches

This chapter attempts to shed some light as to how arches can be designed from first principles, before relying on full computational solutions. Arches take two forms: parabolic or circular. Parabolic are the most common because they do not develop moments when subjected to uniformly distributed loads, as illustrated by Figure 4.1a. In large stadiums and bridges, the dead load is the dominant load, so this attribute leads to efficient designs. In comparison, circular arches do not develop moments when subjected to a uniform load applied normal to the tangent of the arch, which is useful for tunnels. This chapter only considers parabolic arches.

Since UDLs develop no moments in parabolic arches, a critical situation involves increased loading to half of the arch only, since this induces bending moments as illustrated by Figure 4.1b. This chapter reviews the basic theory of arch buckling and explains how to calculate forces and moments due to the loading shown in Figure 4.1c. Worked examples subsequently explain how to estimate the FoS against in-plane buckling. The design method presented explains how to calculate the elastic critical buckling force for parabolic arches using the basic Gordon–Rankine approximation to convert the elastic critical and crushing forces into a design value of buckling strength. This provides 'back of the envelope' estimates of strength for initial design or checking purposes, although full designs will require codes of practice and the use of finite element analysis.

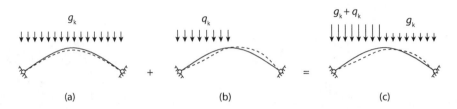

Figure 4.1 Load combination that induces compression and bending. (a) Dead load, (b) imposed load (to half span only) and (c) combined loading.

4.1 ELASTIC CRITICAL BUCKLING AND EFFECTIVE LENGTH

In-plane buckling. Figure 4.2 shows the two main modes of in-plane buckling. Antisymmetric buckling is the most common and is the only mode considered in this chapter, whereas the symmetric or 'snap-through' mode only becomes of concern in shallow arches, which are rarely used in practice.

The famous book *Theory of Elastic Stability* by Timoshenko and Gere (1961) presents a method for calculating the elastic critical buckling load for arches. This book uses the less accurate *equivalent strut method*, as illustrated in Figure 4.3. In this easily understood method, the swept length (S) is used to determine the length (L_{cr}) of a straight strut that would buckle at approximately the same load. Both this and the Timoshenko methods make no allowance for yielding; therefore, the elastic critical buckling force (N_{cr}) is not a real strength but a mathematical one. In the equivalent strut method, Euler's buckling equation is used, i.e.,

$$N_{cr} = \frac{\pi^2 EI}{L_{cr}^2} \tag{4.1}$$

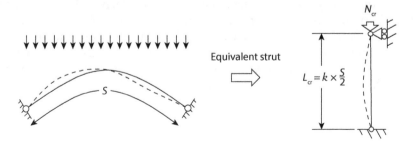

Figure 4.2 Buckling types for the in-plane buckling of parabolic arches. (a) Anti-symmetric buckling and (b) symmetric 'snap-through' buckling.

Figure 4.3 Illustration of the equivalent strut concept for arches.

where E and I are Young's modulus and second moment of area, respectively, and L_{cr} is equivalent to the effective length, which is the swept distance between points of contraflexure.

For antisymmetric buckling, the effective length is

$$L_{cr} = k \times \frac{S}{2} \tag{4.2}$$

where
    k is the effective length factor, taken as 1.0 for two- or three-pinned arches and 0.7 for arches with clamped supports (see Figure 4.4).
    S is the length swept by the arch (see Figure 4.3).

For a parabolic arch, the swept length is

$$S = \frac{L^2}{8f} \ln\left(\frac{4f}{L} + \sqrt{1 + \frac{16f^2}{L^2}} \right) + \frac{L}{2}\sqrt{1 + \frac{16f^2}{L^2}} \tag{4.3}$$

where L is the span and f is the rise as illustrated in Figure 4.5. The x, y coordinates at any point in the parabola are

$$y = \frac{4f}{L^2}\left(Lx - x^2 \right) \tag{4.4}$$

Out-of-plane buckling. If an arch is free to move sideways when viewed from above (on-plan), then out-of-plane buckling must be considered. Arches used for buildings are often restrained against out-of-plane buckling by the roof; however, arches used for bridges often have less restraint. Figure 4.6 shows some common out-of-plane buckling modes and the associated effective lengths. These are the theoretical lower-bound values, although in practice the effective lengths will be longer if connections are not fully rigid.

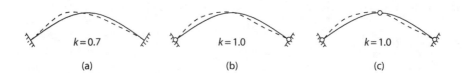

Figure 4.4 Effective length factors. (a) Clamped supports, (b) two-pinned arch and (c) three-pinned arch.

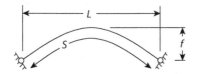

Figure 4.5 Notation and design loading.

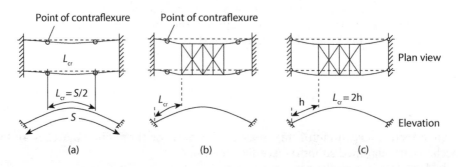

Figure 4.6 Common out-of-plane buckling modes and associated effective lengths. (a) Rigid supports on plan, (b) rigid supports with top bracing and (c) pinned supports with top bracing.

4.2 APPLIED FORCES AND MOMENTS

The only load combination considered in this chapter is dead load over the full span and imposed load over the half span (see Figure 4.5). This provides adverse conditions because the moments induced by the imposed load unbalance the arch, inducing sideways sway, which in turn promotes buckling. From a buckling perspective, the axial force and moment at the quarter span points are of most interest during the initial design, since these points are the locations of the centre of the buckle during antisymmetric buckling as illustrated in Figure 4.2a.

The design moment and axial force at the quarter span points for the loading sketched in Figure 4.5 are

$$M_{L/4} = \frac{q_k L^2}{64} \tag{4.5}$$

$$N_{L/4} = \frac{g_k L^2}{8f}\sqrt{1+\left(\frac{2f}{L}\right)^2} + \frac{q_k L^2}{16f}\sqrt{1+\left(\frac{2f}{L}\right)^2} \tag{4.6}$$

These equations work for both two-pinned and three-pinned arches because the crown of the arch is a point of contraflexure. The support reactions shown in Figure 4.7 are

$$R_{A,V} = \frac{1}{2}g_k L + \frac{3}{8}q_k L$$

$$R_{B,V} = \frac{1}{2}g_k L + \frac{1}{8}q_k L$$

$$R_{A,H} = R_{B,H} = \frac{g_k L^2}{8f} + \frac{q_k L^2}{16f}$$

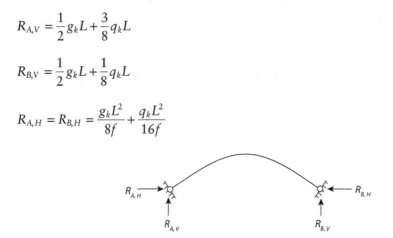

Figure 4.7 Support reactions.

4.3 HOLLOW SECTION OR I- AND H-SECTION ARCHES

The elastic critical buckling force, N_{cr}, tends to infinity at low values of slenderness (see Figure 4.8).

In practice, yielding will cap the strength at $N_{pl, Rd}$ shown in Figure 4.8, which is given as the cross-sectional area multiplied by the yield stress; thus, the hatched region shown in Figure 4.8 represents the theoretical upper limit on strength. Shrinkage stresses from welding or rolling will reduce the strength further. The simplest way to convert N_{cr} to a design value inclusive of imperfections is to use the Gordon–Rankine approximate method presented in Chapter 3. Using this approximation, the buckling force in the absence of applied moments ($N_{b, Rd}$) for a parabolic arch is given by Equation 3.4, i.e.,

$$N_{b, Rd} = \left(\frac{1}{N_{pl, Rd}} + \frac{1}{N_{cr}} \right)^{-1}$$

where $N_{pl, Rd}$ is given by Equation 3.3 and N_{cr} by Equation 4.1.

When checking compression and bending combined, the standard interaction equation presented in Chapter 3 can be used. The critical cross section is at the quarter span point and assuming that out-of-plane moments are zero, Equation 3.6 becomes

$$\frac{N_{1/4}}{N_{b, Rd}} + \frac{\alpha M_{1/4}}{M_{el}} \leq 1 \tag{4.7}$$

where α is the amplification of moments factor given by Equation 3.5 and M_{el} is the elastic moment capacity for the axis of buckling.

If the arch is restrained against lateral movements, as would be the case for most roof arches, then I- or H-sections would be appropriate. Arches that are unrestrained against out-of-plane buckling, such as most arches used for bridges, could also exhibit lateral torsional buckling in addition to the out-of-plane buckling. In which case the stability issues will be complex, and for this reason hollow-section members, such as square hollow sections, should be used where out-of-plane buckling can occur. This is because hollow-section members are not normally prone to lateral torsional buckling if the depth is not significantly greater than the width.

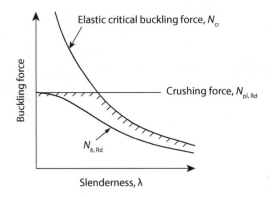

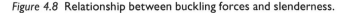

Figure 4.8 Relationship between buckling forces and slenderness.

The interaction equation (Equation 4.7) should also not be used with Class 4 cross sections, since these will fail by local buckling before attaining their elastic bending strength. This is not an issue for any of the standard hot-rolled sections available from the steel mills, since these sections are almost all class 1 or 2.

Example 4.1: Buckling of a simple arch

A two-pinned parabolic arch spans 25 m and has a rise of 9 m. It is restrained against out-of-plane buckling, which is about the vertical z–z axes shown in Figure 4.9. Determine if the arch is likely to exhibit antisymmetric (in-plane) buckling when resisting the ULS loads shown. Young's modulus is 210,000 N/mm² and the yield stress is 335 N/mm².

Step 1: Calculate the design moment and thrust at the quarter span points.

From Equation 4.5, the moment at the quarter span points is

$$M_{L/4} = \frac{q_k L^2}{64}$$

$$M_{1/4} = \frac{100 \times 25^2}{64} = 977 \text{ kN.m}$$

and the axial force from Equation 4.6 is

$$N_{L/4} = \frac{g_k L^2}{8f} \sqrt{1 + \left(\frac{2f}{L}\right)^2} + \frac{q_k L^2}{16f} \sqrt{1 + \left(\frac{2f}{L}\right)^2}$$

$$N_{1/4} = \frac{300 \times 25^2}{8 \times 9} \sqrt{1 + \left(\frac{2 \times 9}{25}\right)^2} + \frac{100 \times 25^2}{16 \times 9} \sqrt{1 + \left(\frac{2 \times 9}{25}\right)^2} = 3744 \text{ kN}$$

Step 2: Calculate the crushing strength of the arch member:

Cross-sectional area $= 800 \times 500 - 720 \times 420 = 97600 \text{ mm}^2$

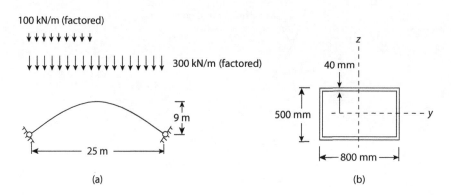

(a)   (b)

Figure 4.9 In-plane buckling of an arch problem. (a) Elevation showing loading and (b) cross section.

and the crushing strength from Equation 3.3 is

$$N_{pl,\,Rd} = f_y \times \text{Area} = 355 \times 97600 \times 10^{-3} = 34648 \text{ kN}$$

Step 3: Estimate the elastic critical buckling strength of the arch.

This example considers only in-plane buckling; therefore, the second moment of area of the cross section about the horizontal $(y-y)$ axes is used in calculating N_{cr}, where

$$I_y = \frac{800 \times 500^3 - 720 \times 420^3}{12} = 3888 \times 10^6 \text{ mm}^4$$

From Equation 4.3, the arc length of the arch is

$$S = \frac{25^2}{8 \times 9} \ln\left(\frac{4 \times 9}{25} + \sqrt{1 + \frac{16 \times 9^2}{25^2}} \right) + \frac{25}{2} \sqrt{1 + \frac{16 \times 9^2}{25^2}} = 32 \text{ m}$$

And from Equation 4.2, the effective length is

$$L_{cr} = 1.0 \times \frac{32}{2} = 16 \text{ m}$$

From Equation 3.1, the elastic critical buckling force is

$$N_{cr} = \frac{\pi^2 EI}{L_{cr}^2} = \frac{\pi^2 \times 210000 \times 3888 \times 10^6}{16000^2} \times 10^{-3} = 31478 \text{ kN}$$

Step 4: Estimate the buckling force in the absence of moments.

From Equation 3.4

$$N_{b,\,Rd} = \left(\frac{1}{N_{pl,\,Rd}} + \frac{1}{N_{cr}} \right)^{-1} = \left(\frac{1}{34648} + \frac{1}{31478} \right)^{-1} = 16494 \text{ kN}$$

Step 5: Calculate the moment capacity of the arch member.

$$M_{el,\,y} = \frac{f_y I}{D/2} = \frac{335 \times 3888 \times 10^6}{250} \times 10^{-6} = 5210 \text{ kN.m}$$

Step 6: Check for combined compression and bending.

From Equation 3.5, the amplification of moments factor is

$$\alpha = \frac{1}{1 - N/N_{cr}} = \frac{1}{1 - 3744/31478} = 1.13$$

Now inputting the above results into the main interaction equation (Equation 4.7)

$$\frac{N_{1/4}}{N_{b,\,Rd}} + \frac{\alpha M_{1/4}}{M_{el}} \le 1$$

$$\frac{3744}{16494} + \frac{1.13 \times 977}{5210} = 0.44$$

Since this is significantly less than 1.0, this would indicate that the arch is unlikely to exhibit antisymmetric buckling.

4.4 LACED GIRDER ARCHES

A method for analysing in-plane buckling of slender trusses was presented in Sections 3.5 and 3.6. This can be adapted to provide approximate solutions for lattice girder arches subjected to in-plane buckling. Equation 3.20 becomes

$$\frac{N_{1/4}}{N_{b,\,Rd}} + \frac{\alpha M_{1/4}}{M_{Rd}} \le 1 \tag{4.8}$$

where $N_{b,\,Rd}$ is calculated using Equation 3.16 and M_{Rd} is the moment capacity of the lattice girder. This is less than the elastic moment capacity, because buckling of the chords will occur at a stress lower than the yield stress. The calculation of this moment capacity is described in Section 3.6.

The effects of deformations in web members will reduce the elastic critical bucking load below that defined by Equation 4.1. If the effects of shear distortions are of concern, then the method for calculating the elastic critical buckling loads for slender lattice girders in compression presented by Timoshenko and Gere (1961) can give an approximate solution (see Equations 3.17 and 3.18), although this often makes little practical difference and has therefore been ignored in the examples presented herein.

Example 4.2: Lattice girder arch buckling

Figure 4.10 shows a lattice girder arch that is restrained against out-of-plane buckling. The yield stress is 355 N/mm², Young's modulus 210,000 N/mm² and the main chord members are 300 mm wide square hollow section (SHS) members with $I = 19,440$ cm⁴ and $A = 142$ cm². The diagonal web members are 200 mm wide SHS members with $A = 38$ cm². Check the ability to resist the ULS (factored) loads shown in Figure 4.10.

Step 1: Calculate the design moment and thrust at the quarter span points.

From Equations 4.5 and 4.6

$$M_{L/4} = \frac{q_k L^2}{64} = \frac{26.6 \times 100^2}{64} = 4156 \text{ kN.m}$$

$$N_{L/4} = \frac{g_k L^2}{8f} \sqrt{1 + \left(\frac{2f}{L}\right)^2} + \frac{q_k L^2}{16f} \sqrt{1 + \left(\frac{2f}{L}\right)^2}$$

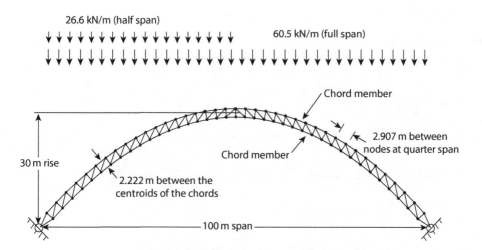

Figure 4.10 Elevation of a lattice girder arch.

$$N_{1/4} = \frac{60.5 \times 100^2}{8 \times 30} \sqrt{1 + \left(\frac{2 \times 30}{100}\right)^2} + \frac{26.6 \times 100^2}{16 \times 30} \sqrt{1 + \left(\frac{2 \times 30}{100}\right)^2} = 3586 \text{ kN}$$

Step 2: Calculate the 'local buckling' strength of the arch (illustrated in Figure 3.17b).

From Equation 3.3, the crushing strength of a single chord member is

$$N_{\text{pl, Rd}} = f_y A = 355 \times 142 \times 10^2 \times 10^{-3} = 5040 \text{ kN}$$

For this local buckling mode, the effective length of the chord members is L_{cr} = 2907 mm (see Figure 4.10), and from Equation 3.1, the elastic critical buckling force is

$$N_{\text{cr}} = \frac{\pi^2 EI}{L_{\text{cr}}^2}$$

$$N_{\text{cr}} = \frac{\pi^2 \times 210000 \times 19440 \times 10^4}{2907^2} \times 10^{-3} = 47680 \text{ kN}$$

And from Equation 3.4, the buckling strength of a single chord member is

$$N_{b, \text{Rd}} = \left(\frac{1}{N_{\text{pl, Rd}}} + \frac{1}{N_{\text{cr}}}\right)^{-1}$$

$$N_{b, \text{Rd}} = \left(\frac{1}{5040} + \frac{1}{47680}\right)^{-1} = 4558 \text{ kN}$$

The force required to cause local buckling is

$$N_{\text{local}} = 2 \times N_{b, \text{Rd}} = 2 \times 4558 = 9116 \text{ kN}$$

Step 3: Estimate the elastic critical buckling strength (Figure 4.3) of the arch.

Figure 4.10 shows that the centroids of the inner and outer chords are 2.222 m apart. From Equation 3.15, the second moment of area of the lattice girder is

$$I = \sum \left(I_{chord} + Area_{chord}.r^2 \right)$$

$$I = 2 \times \left(19440 \times 10^4 + 142 \times 10^2 \times (2222/2)^2 \right) = 3.54 \times 10^{10} \, mm^4$$

From Equation 4.3, the arc length $(S) = 120.4$ m and from Equation 4.2,

$$L_{cr} = 1.0 \times \frac{120.4}{2} = 60.2 \, m$$

And the elastic critical buckling force from Equation 4.1 is

$$N_{cr} = \frac{\pi^2 \times 210000 \times 3.54 \times 10^{10}}{60200^2} \times 10^{-3} = 20245 \, kN$$

Step 4: Estimate the buckling force in the absence of moments.

From Equation 3.16, the axial force at the quarter span points that is likely to cause anti-symmetric buckling (in the absence of moments) as illustrated in Figure 4.2a is

$$N_{b, Rd} = \left(\frac{1}{N_{local}} + \frac{1}{N_{cr}} \right)^{-1}$$

$$N_{b, Rd} = \left(\frac{1}{9116} + \frac{1}{20245} \right)^{-1} = 6286 \, kN$$

Step 5: Calculate the moment capacity of the arch.

From Equation 3.13

$$M = N \times z$$

$$M_{Rd} = 4558 \times 2.222 = 10128 \, kN.m$$

Step 6: Check for combined compression and bending.

From Equation 3.5, the moment amplification factor is

$$\alpha = \frac{1}{1 - \dfrac{N}{N_{cr}}}$$

$$\alpha = \frac{1}{1 - \dfrac{3586}{20245}} = 1.215$$

Finally, the main interaction equation that combines compression and bending is solved (Equation 4.7)

$$\frac{N_{1/4}}{N_{b,\,Rd}} + \frac{\alpha M_{1/4}}{M_{Rd}} \leq 1$$

$$\frac{3586}{6286} + \frac{1.215 \times 4156}{10128} = 1.07$$

Since this is greater than 1.0, the arch is not strong enough. A finite element analysis (FEA) solution showed that buckling occurred under the loads shown in Figure 4.10. However, the FEA solution will overestimate strength, because it does not include residual stresses, amongst some other possible imperfections.

A final consideration is that both the inner and outer chords of the arch need to be restrained against out-of-plane buckling. Leaving either free will result in a collapse.

Example 4.3: Lattice girder arch buckling

Figure 4.11 shows the FEA predicted buckling shape and buckling loads for a long-span arch. The arch is restrained against out-of-plane buckling and has similar proportions to the arch in Figure 4.12. Check if it can resist the loads shown in Figure 4.11.

Basic data. The cross-sectional dimensions are shown in Figure 4.13. The 12 individual struts each have a diameter of 457 mm, wall thickness of 40 mm, cross-sectional area of 524 cm² and I of 114,900 cm⁴. The yield stress is 345 N/mm² and E is 210,000 N/mm².

The solution to this problem is broken down into five separate steps.

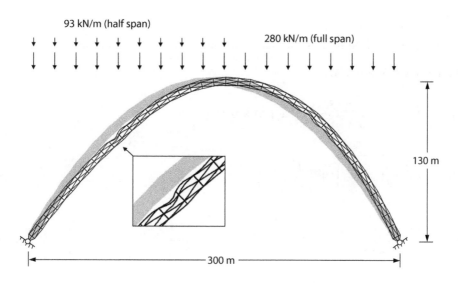

Figure 4.11 Arch proportions, loading and buckling shape.

Figure 4.12 Wembley Arch, London.

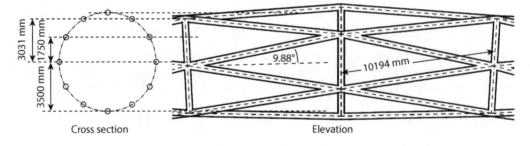

Cross section   Elevation

Figure 4.13 Details of the arch.

Step 1: Calculate the design moment and thrust at the quarter span points.

From Equations 4.5 and 4.6,

$$M_{1/4} = \frac{q_k L^2}{64} = \frac{93 \times 300^2}{64} \times 10^{-3} = 130.8 \text{ MN.m}$$

$$N_{L/4} = \frac{g_k L^2}{8f} \sqrt{1 + \left(\frac{2f}{L}\right)^2} + \frac{q_k L^2}{16f} \sqrt{1 + \left(\frac{2f}{L}\right)^2}$$

$$N_{1/4} = \frac{280 \times 10^{-3} \times 300^2}{8 \times 130} \sqrt{1 + \left(\frac{2 \times 130}{300}\right)^2} + \frac{88 \times 10^{-3} \times 300^2}{16 \times 130} \sqrt{1 + \left(\frac{2 \times 130}{300}\right)^2} = 37.1 \text{ MN}$$

Step 2: Calculate the local buckling strength of the arch.

The first part of this process is to determine the crushing capacity of each strut using Equation 3.3:

$$N_{pl, Rd} = f_y A = 345 \times 524 \times 10^2 \times 10^{-6} = 18.08 \text{ MN}$$

The effective length for this local buckling mode, L_{cr} = 1.0 L, where from Figure 4.13 L = 10,194 mm.

From Equation 3.1,

$$N_{cr} = \frac{\pi^2 \times 210000 \times 114900 \times 10^4}{10194^2} \times 10^{-6} = 22.92 \text{ MN}$$

and from Equation 3.4, the compression strength of each strut in the truss is

$$N_{b, Rd} = \left(\frac{1}{18.08} + \frac{1}{22.92} \right)^{-1} = 10.11 \text{ MN}$$

The struts are inclined to the centre line of the arch by 9.88° (see Figure 4.13); therefore, the resistance along the axis of the arch is

$$N_{b, Rd} = 10.11 \times \cos 9.88 = 9.96 \text{ MN}$$

The truss comprises 12 separate leg members; therefore, the local compression strength is

$$N_{local} = 12 \times N_{b, Rd} = 12 \times 9.96 = 119.5 \text{ MN}$$

Step 3: Estimate the elastic critical buckling strength (Figure 4.3) of the arch.

From Equation 3.15, the second moment of area of the lattice girder is

$$I = 12 \times 114900 \times 10^4 + 524 \times 10^2 \times (2 \times 3500^2 + 4 \times 3031^2 + 4 \times 1750^2) = 3.865 \times 10^{12} \text{ mm}^4$$

Because the arch struts are inclined to the centre line by an angle of 9.88°, the effective second moment of area is reduced slightly to

$$I = 3.865 \times 10^{12} \times \cos 9.88 = 3.808 \times 10^{12} \text{ mm}^4$$

From Equation 4.3, the arc length (S) is 414 m. From Equation 4.2, the effective length is 207 m and

$$N_{cr} = \frac{\pi^2 EI}{L_{cr}^2}$$

$$N_{cr} = \frac{\pi^2 \times 210000 \times 3.808 \times 10^{12}}{207000^2} \times 10^{-6} = 184.2 \text{ MN}$$

Step 4: Estimate the buckling force in the absence of moments.

From Equation 3.16, the buckling force in the absence of moments is

$$N_{b,\,Rd} = \left(\frac{1}{N_{local}} + \frac{1}{N_{cr}} \right)^{-1}$$

$$N_{b,\,Rd} = \left(\frac{1}{119.5} + \frac{1}{184.2} \right)^{-1} = 72.5 \text{ MN}$$

Step 5: Calculate the moment capacity of the arch.

The bending stress, which will cause buckling of the arch members, is the strut capacity divided by the area, i.e.,

$$\text{Buckling stress, } \sigma = \frac{N_{b,\,Rd}}{A} = \frac{9.96 \times 10^6}{524 \times 10^2} = 190 \text{ N/mm}^2$$

From the engineer's beam equation, the corresponding moment is

$$M_{Rd} = \frac{\sigma \times I}{y} = \frac{190 \times 3.808 \times 10^{12}}{3500} \times 10^{-9} = 206.7 \text{ MN.m}$$

Step 6: Check for combined compression and bending.

From Equation 3.5, the amplification factor is

$$\alpha = \frac{1}{1 - \dfrac{N}{N_{cr}}}$$

$$\alpha = \frac{1}{1 - \dfrac{37.1}{184.2}} = 1.252$$

Finally, Equation 4.8 is solved

$$\frac{N_{1/4}}{N_{b,\,Rd}} + \frac{\alpha M_{1/4}}{M_{Rd}} \le 1$$

$$\frac{37.1}{72.5} + \frac{1.252 \times 130.8}{206.7} = 1.30$$

Therefore, this method predicts the arch will certainly buckle under the applied loads. A finite element analysis of this problem predicted buckling at the exact loads shown in Figure 4.11, although as stated by Allan Mann in his foreword to this book:

"Indeed any presumption that computer output is 'accurate' is itself a fiction."

This is certainly true in this case, since the computational solution did not include residual stresses.

4.5 CALCULATION OF ELASTIC CRITICAL BUCKLING LOAD USING THE TIMOSHENKO METHOD

The *equivalent strut method* is easy to understand, although the method reported in *Theory of Elastic Stability* by Timoshenko and Gere (1961) is more accurate. The Timoshenko elastic critical buckling load is a mathematical solution and it should correspond exactly with the results from an elastic FEA solution. According to Timoshenko, the elastic critical load is

$$g_{cr} = \gamma \frac{EI}{L^3} \tag{4.9}$$

Table 4.1 shows the arch buckling factor (γ) for parabolic arches resisting pure UDLs. The corresponding (elastic critical buckling) force at the quarter span point (N_{cr}) is calculated by inputting g_{cr} into Equation 4.6.

It must be remembered that real structures buckle at loads far below the elastic critical buckling load, because imperfections, such as yielding, are not included. Timoshenko and Gere did not go beyond elastic theory, although the Gordon–Rankine approach explained previously can be used to estimate the strength.

Example 4.4: Repeat Example 4.1 using the Timoshenko method

Steps 1, 2 and 5 are identical to Example 4.1.

Step 3: Estimate the elastic critical buckling strength of the arch

The f/L ratio of the arch is

$$\frac{f}{L} = \frac{9}{25} = 0.36$$

From Table 4.1, $\gamma = 44.94$ (found by linear interpolation) and from Equation 4.9,

$$g_{cr} = 44.94 \times \frac{210000 \times 10^3 \times 3888 \times 10^{-6}}{16^3} = 2348 \text{ kN/m}$$

From Equation 4.6, this will produce a force at the quarter span point of

$$N_{cr} = \frac{2348 \times 25^2}{8 \times 9} \sqrt{1 + \left(\frac{2 \times 9}{25}\right)^2} = 25115 \text{ kN}$$

Table 4.1 Elastic critical buckling factors (γ)

$\dfrac{f}{L}$	No hinges (fixed supports)	Two hinges	Three hinges
0.1	60.7	28.5	22.5
0.2	101.0	45.4	39.6
0.3	115.0	46.5	46.5
0.4	111.0	43.9	43.9
0.5	97.4	38.4	38.4

Source: Adapted from Timoshenko and Gere (1961).

Step 4: Estimate the buckling force in the absence of moments.

From Equation 3.4,

$$N_{b,\,Rd} = \left(\frac{1}{34648} + \frac{1}{25115} \right)^{-1} = 14561\,\text{kN}$$

Step 6: Check for combined compression and bending.

From Equation 3.5,

$$\alpha = \frac{1}{1 - 3744/25115} = 1.175$$

From Equation 4.7,

$$\frac{3744}{14561} + \frac{1.175 \times 977}{5210} = 0.48$$

This utilisation factor is slightly higher than that found in Example 4.1 (0.44).

Problems

Solutions to these problems are provided at https://www.crcpress.com/9781498741217

P.4.1. A two-pinned arch spans 22.5 m and has a rise of 11 m (Figure 4.14). It is restrained against out-of-plane buckling, the yield stress is 345 N/mm² and E = 210,000 N/mm².

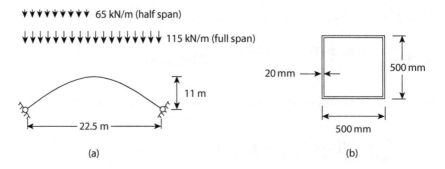

Figure 4.14 In-plane buckling of an arch for P4.1. (a) Elevation showing loading and (b) cross section.

a. Determine the moment and thrust at the quarter span points.
b. Determine the crushing strength of the arch.
c. Estimate the elastic critical buckling strength of the arch.
d. Determine if the arch is likely to exhibit antisymmetric (in-plane) buckling under the applied loads.

Ans. (a) $M_{1/4}$ = 514.2 kN.m, $N_{1/4}$ = 1187 kN, (b) 13,248 kN, (c) 11,244 kN and (d) No, 0.48 < 1.

P.4.2. A two-pinned arch spans 35 m and has a rise of 17 m (see Figure 4.15). It is restrained against out-of-plane buckling, the yield stress is 345 N/mm² and E = 210,000 N/mm².

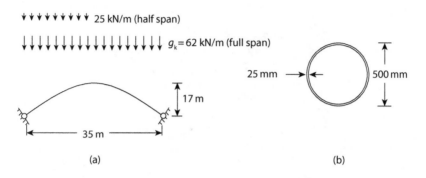

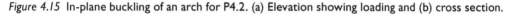

Figure 4.15 In-plane buckling of an arch for P4.2. (a) Elevation showing loading and (b) cross section.

a. Determine the moment and thrust at quarter span points.
b. Determine the crushing strength.
c. Estimate the elastic critical buckling strength.
d. Determine the FoS against buckling under these loads.

Ans. (a) $M_{1/4}$ = 478.5 kN.m, $N_{1/4}$ = 935.5 kN, (b) 12,871 N, (c) 3363 kN and (d) 1.23.

REFERENCE

Timoshenko, S. P. and Gere, M., 1961. *Theory of Elastic Stability.* 2nd Edition. New York: McGraw-Hill.

Chapter 5

Buckling of thin-walled structures

The basic theory governing the design of box girder bridges also applies to other *thin-walled structures*, such as ships, which are built up from thin plates stiffened to increase strength. Figure 5.1 shows the main types of buckling, namely compression buckling and shear buckling, both of which occur in stiffened plates, such as that shown in Figure 5.2.

Strength is governed by the theoretical values of the elastic critical buckling compression stress (σ_{cr}) and shear stress (τ_{cr}). These do not account for yielding; consequently, the elastic critical buckling stress can be many times higher than the yield stress because it is a mathematical limit rather than a real one. The hatched boundaries shown in Figure 5.3 provide a basic starting point for defining the upper limit on strength as a function of plate slenderness. In practice, imperfections, such as weld shrinkage stresses, lead to buckling at compression stresses somewhat below the hatched boundary shown in Figure 5.3a.

Thin-walled structures can have a 'post-buckling' reserve of strength, meaning that the failure stress can, in some circumstances, be higher than the elastic critical value. This is true for shear buckling of stiffened plates. Bridge engineers can utilise this strength, which is known as *tension field action* (Figure 5.4a). However, plate stiffeners require strengthening in order to resist the forces developed due to the truss action that forms (Figure 5.4b) and the design becomes complex. In the interests of simplicity, tension field action is not covered in this chapter.

Very wide plates are often used for box girders. These are stiffened in order to raise the buckling stress as shown in Figure 5.2. For an efficient design, engineers aim to raise the buckling stress close to the yield stress. This is called a *fully stressed design*. Bridge codes contain

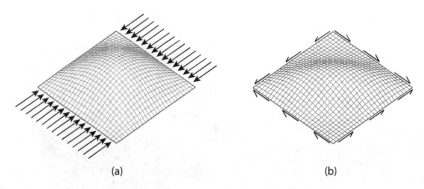

Figure 5.1 Buckling of plates supported on all four edges. (a) Compression buckling and (b) shear buckling.

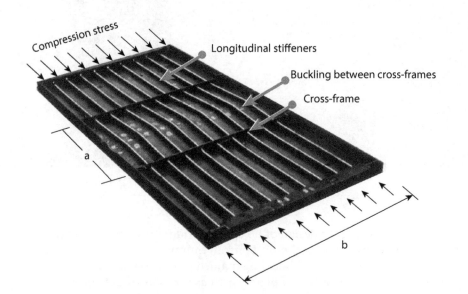

Figure 5.2 Buckling of a stiffened panel.

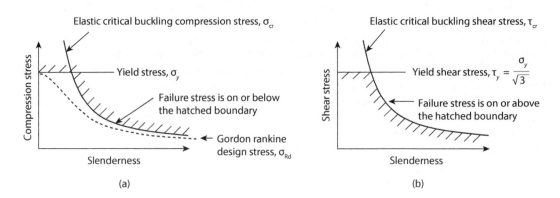

Figure 5.3 Relationship between slenderness and buckling stress for plates. (a) Compression buckling and (b) shear buckling.

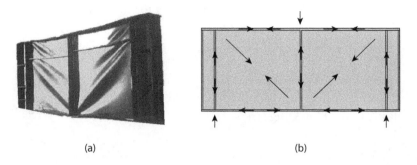

(a)   (b)

Figure 5.4 Post-buckling reserve of strength known as *tension field action*. (a) Post-elastic buckling and (b) the truss model.

complex methods, although all bridge codes rely on the use of elastic critical buckling stresses, with modifications to account for imperfections, such as yielding and the compression stresses induced from the shrinkage of welds after cooling. The theory presented in this chapter is very much simplified and uses a simple Gordon–Rankine approach in order to convert elastic critical stresses to design stresses (see Figure 5.3a). This provides a quick back-of-the-envelope estimate of strength. It also provides a sanity check for the output from more complex methods. Linear elastic finite element analysis is used for the design of bridges, which should in theory provide buckling stresses equal to the elastic critical buckling stresses determined from plate buckling theory. Therefore, the elastic critical buckling stresses provide a useful reference point. Further simplification is achieved by omitting the partial safety factors for resistance, which are assumed to be equal to 1.0 throughout and therefore not shown in the design expressions.

5.1 UNSTIFFENED PLATES IN COMPRESSION

Figure 5.5a shows buckling of a plate free along two edges. When loaded in axial compression, the plate will want to expand in the transverse direction. If the supports restrain this sideways movement, lateral stress will develop. According to Hooke's law (1678), the strain in an element subjected to stresses in both the x and y directions (σ_x and σ_y) is

$$\varepsilon_x = \frac{1}{E}\left(\sigma_x - \nu\sigma_y\right)$$

where
   ν is Poisson's ratio.
   E is Young's modulus.

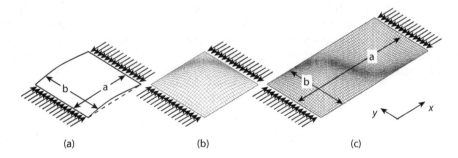

(a)   (b)   (c)

Figure 5.5 Coordinate system and buckling modes. (a) Free on two edges, (b) simply supported on four edges and (c) simply supported on four edges.

It can be shown that the elastic critical buckling stress for the plate shown in Figure 5.5a is

$$\sigma_{cr} = \frac{E\pi^2}{12(1-v^2)}\left(\frac{t}{a}\right)^2 \tag{5.1}$$

where
    σ_{cr} is the elastic critical buckling stress.
    t is the plate thickness.
    a is the plate length.

Equation 5.1 is the same as the Euler elastic critical buckling stress equation for a simply supported rectangular strut (Equation 3.1), *except* for the term $1-v^2$, which has the effect of increasing σ_{cr} by 10% for steel plates. For the more useful case of a plate simply supported on all four edges (Figure 5.5b and c), the solution becomes

$$\sigma_{cr} = k\frac{E\pi^2}{12(1-v^2)}\left(\frac{t}{b}\right)^2 \tag{5.2}$$

where
    k is the buckling coefficient.
    b is the plate width (see Figure 5.5c).
    t is the plate thickness.

For steel, $v = 0.3$, $E = 210,000$ N/mm² and Equation 5.2 becomes

$$\sigma_{cr} = k\frac{210000\pi^2}{12(1-0.3^2)}\left(\frac{t}{b}\right)^2$$

which solves to

$$\sigma_{cr} = k \times 190000 \times \left(\frac{t}{b}\right)^2 \text{ N/mm}^2 \tag{5.3}$$

It is important to realise that the buckling stress for a plate free on two edges (Figure 5.5a) is dependent on the length (a), *whereas* for a plate restrained on four edges (Figure 5.5b and c), it is dependent on width (b). This is illustrated by Figure 5.5c, which shows a plate twice as long as that in Figure 5.5b. Since the width of both plates is identical, the elastic critical buckling stress is unaffected despite the increased length. This dependency on width rather than length means that longitudinal stiffeners are far more effective than transverse ones.

The factor k in Equation 5.3 is known as the *buckling coefficient*, which for a plate simply supported on all four edges and loaded in uniform compression (Figure 5.5b and c) is never less than 4.0. Figure 5.6 itemises buckling coefficients for a range of different plate types and loadings. For example, Figure 5.6c shows that k equals 23.9, for a plate simply supported on four edges and subjected to a bending stress distribution.

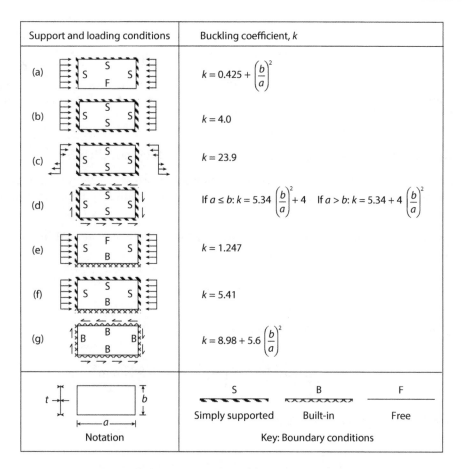

Figure 5.6 Buckling coefficients for unstiffened plates.

Design strength. The elastic critical buckling compression stresses define the theoretical upper limit on strength, as illustrated in Figure 5.3a. In practice, yielding and weld shrinkage stresses, as well as lack of straightness, all combine to reduce strength. The codes of practice have a variety of semi-empirical methods to account for these defects. These can be complex, but an approximation for the design value of buckling stress (σ_{Rd}) can be made using a Gordon–Rankine type approximation, i.e.,

$$\frac{1}{\sigma_{Rd}} = \frac{1}{\sigma_y} + \frac{1}{\sigma_{cr}}$$

which rearranges to

$$\sigma_{Rd} = \left(\frac{1}{\sigma_y} + \frac{1}{\sigma_{cr}} \right)^{-1} \tag{5.4}$$

where

    σ_{Rd} is the design value for the buckling stress.
    σ_y is the yield stress.

Design procedure. The first step is to identify the appropriate buckling coefficient from Figure 5.6, which applies to the particular plate concerned. The plate length (a) and width (b) are defined by the distance between plates connected at right angles. The boundaries are usually either free or simply supported. The next step is to determine the elastic critical buckling stress and, finally, the design value of the buckling stress, using Equation 5.4. An efficient design will achieve buckling stresses close to the yield value. If this is not the case, then stiffeners are added.

Example 5.1: Plate buckling with different boundary conditions

A 20 mm thick steel plate is subjected to compression stress as shown in Figure 5.7. Determine the design buckling stresses for each of the two boundary conditions shown if the yield stress is 265 N/mm².

For the boundary conditions in Figure 5.7a, the buckling coefficient shown in Figure 5.6a applies; therefore, the buckling coefficient

$$k = 0.425 + (b/a)^2 = 0.425 + (1200/3000)^2 = 0.585$$

From Equation 5.3, the elastic critical buckling stress is

$$\sigma_{cr} = k \times 190000 \times \left(\frac{t}{b}\right)^2 = 0.585 \times 190000 \left(\frac{20}{1200}\right)^2 = 30.9 \text{ N/mm}^2$$

From Equation 5.4, the design stress is

$$\sigma_{Rd} = \left(\frac{1}{\sigma_y} + \frac{1}{\sigma_{cr}}\right)^{-1} = \left(\frac{1}{265} + \frac{1}{30.9}\right)^{-1} = 27.7 \text{ N/mm}^2$$

Now considering the boundary conditions in Figure 5.7b, the buckling coefficient shown in Figure 5.6b applies (i.e. k = 4.0). Therefore,

$$\sigma_{cr} = 4.0 \times 190000 \left(\frac{20}{1200}\right)^2 = 211.1 \text{ N/mm}^2$$

and the buckling stress is

$$\sigma_{Rd} = \left(\frac{1}{265} + \frac{1}{211.1}\right)^{-1} = 117.5 \text{ N/mm}^2$$

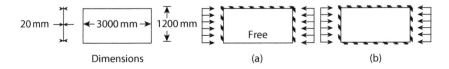

Figure 5.7 Two plates with different boundary conditions.

It can be seen that adding the additional boundary support has increased the design stress from 27.7 N/mm² to 117.5 N/mm². This is because the wavelength of the buckle is reduced when the edge restraint is added.

5.2 SHEAR BUCKLING OF UNSTIFFENED PLATES

Plate girder webs are usually very slender and fail by shear buckling rather than yielding. To control this, intermediate web stiffeners are used to raise the shear strength, as illustrated in Figure 5.8. Buckling theory tells us that

$$\tau_{cr} = k \frac{\pi^2 E}{12(1-\upsilon^2)} \left(\frac{t}{b}\right)^2 \tag{5.5}$$

where
  τ_{cr} is the elastic critical buckling shear stress.
  k is the buckling coefficient.
  b is the plate width.
  t is plate thickness.

For steel, Poisson's ratio $\upsilon = 0.3$ and Young's modulus $E = 210,000$ N/mm² and Equation 5.5 becomes

$$\tau_{cr} = k \times 190000 \times \left(\frac{t}{b}\right)^2 \tag{5.6}$$

For the most common situation of plates simply supported on all four edges (Figure 5.6d), the buckling coefficient is

$$k = 5.34 + 4\left(\frac{b}{a}\right)^2 \quad \text{if } a > b \tag{5.7}$$

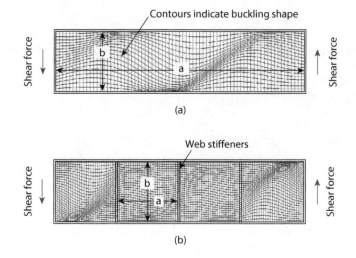

Figure 5.8 Illustration of shear buckling for a plate girder web. (a) No intermediate web stiffeners and (b) with intermediate stiffeners.

$$k = 5.34\left(\frac{b}{a}\right)^2 + 4 \quad \text{if } a \le b \tag{5.8}$$

where a is the length and b is the width as illustrated in Figure 5.8. As stated previously, plates bounded on four sides by flanges and web stiffeners have a post-buckling reserve of strength due to a phenomenon called *tension field action*, which is illustrated in Figure 5.4. This means that the elastic critical shear stress generally provides a conservative estimate of strength, despite imperfections, such as weld shrinkage stresses and lack of straightness. For this reason, the design stress can be taken as either the elastic critical shear stress or the yield shear stress, whichever is less. The shear stress, which will cause first yielding (Equation 2.2), is

$$\tau_y = \frac{\sigma_y}{\sqrt{3}} \tag{5.9}$$

The design value of shear stress (τ_{Rd}) is the lesser of either τ_{cr} or τ_y, i.e.,

$$\tau_{Rd} = \tau_{cr} \text{ or } \frac{\sigma_y}{\sqrt{3}} \text{ (whichever is less)} \tag{5.10}$$

and the design shear strength (V_{Rd}) is approximately given by

$$V_{Rd} = \tau_{Rd} \times \text{plate area} \tag{5.11}$$

Example 5.2: Shear buckling in a plate girder

A plate girder web is 2100 mm deep and 20 mm thick. The steel has a yield stress of 265 N/mm² and web stiffeners are located at 8000 mm centres (see Figure 5.8a).

1. Determine the shear strength.
2. Determine what effect adding three intermediate web stiffeners at 2000 mm centres will have on the shear strength, as shown in Figure 5.8b.

1. Without intermediate web stiffeners, $a = 8000$ mm; therefore, $a > b$ and from Equation 5.7

$$k = 5.34 + 4\left(\frac{b}{a}\right)^2 = 5.34 + 4\left(\frac{2100}{8000}\right)^2 = 5.616$$

The elastic critical buckling stress from Equation 5.6 is

$$\tau_{cr} = k \times 190000 \times \left(\frac{t}{b}\right)^2 = 5.616 \times 190000 \times \left(\frac{20}{2100}\right)^2 = 96.8 \text{ N/mm}^2$$

Because the yield stress is 265 N/mm², the yield shear stress from Equation 5.9 is

$$\tau_y = \frac{265}{\sqrt{3}} = 153 \text{ N/mm}^2$$

Since $\tau_{cr} < \tau_y$, elastic buckling rather than yielding will initiate failure. The design shear strength (V_{Rd}) is the design shear stress multiplied by the web plate cross-sectional area from Equation 5.11

$$V_{Rd} = 2100 \times 20 \times 96.8 \times 10^{-3} = 4065 \text{ kN}$$

2. The introduction of intermediate web stiffeners reduces the web length (a) to 2000 mm. Now $a < b$ and from Equation 5.8

$$k = 5.34 \left(\frac{2100}{2000} \right)^2 + 4 = 9.89$$

and from Equation 5.6

$$\tau_{cr} = 9.89 \times 190000 \times \left(\frac{20}{2100} \right)^2 = 170.4 \text{ N/mm}^2$$

Since $\tau_{cr} > \tau_y$, failure will occur by yielding rather than elastic buckling and the design shear strength is

$$V_{Rd} = 2100 \times 20 \times 153 \times 10^{-3} = 6426 \text{ kN}$$

To summarise, the introduction of three intermediate web stiffeners has increased the shear strength from 4065 kN to 6426 kN. The addition of these stiffeners also has the advantage of keeping the web straight, since distortion due to weld shrinkage stresses may occur otherwise.

5.3 UNSTIFFENED PLATES IN COMPRESSION AND SHEAR

It is common for high shear forces to coincide with high moments. Figure 5.9 shows buckling from shear, bending and combined shear and bending stresses. Engineers often deal with combined loading problems by using 'interaction equations' in which the ratios of applied forces and strengths are summed, with the objective being to ensure that the total sum of these ratios is less than 1.0. A variation of this approach can be used for plate buckling.

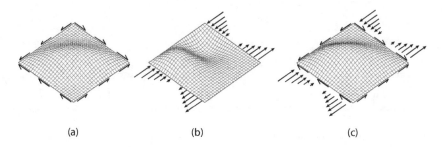

(a)   (b)   (c)

Figure 5.9 Buckling shapes for shear, bending and shear + bending combined. (a) Shear stresses, (b) bending stresses and (c) shear + bending stresses.

Shear combined with bending. Plates subjected to shear and bending (Figure 5.9c) can be assessed using the following interaction equation:

$$\left(\frac{\sigma_{Ed}}{\sigma_{Rd}}\right)^2 + \left(\frac{\tau_{Ed}}{\tau_{Rd}}\right)^2 \leq 1.0 \qquad (5.12)$$

where
  σ_{Ed} is the applied compression stress in the web developed by bending moments.
  τ_{Ed} is the applied shear stress.
  and
  σ_{Rd} is the design stress for a plate resisting bending stresses (not σ_{cr}).
  τ_{Rd} is design shear stress (not τ_{cr}).

Shear combined with compression. In plates subjected to compression and shear, the following interaction equation should be satisfied:

$$\frac{\sigma_{Ed}}{\sigma_{Rd}} + \left(\frac{\tau_{Ed}}{\tau_{Rd}}\right)^2 \leq 1.0 \qquad (5.13)$$

Both Equations 5.12 and 5.13 from Allen & Bulson (1980) are semi-empirical, i.e., the evidence that they work is partly based on observations and partly on theory.

Example 5.3: Plate girder with web stiffeners

Figures 5.10 and 5.11 show a plate girder that resists a 200 kN/m UDL inclusive of self-weight. It is restrained against sideways movement so that lateral torsional buckling is not possible and the web is stiffened by 30 mm thick stiffeners positioned at 1.25 m centres.

1. Determine the support reactions and sketch the bending moment and shear force diagrams.
2. If the yield stress is 265 N/mm², determine if the web stiffeners can resist the concentrated force of the support reaction at Support A without buckling.
3. Determine if the web is strong enough to resist the combined shear force and moments at Support A.
4. Determine if the flanges can resist the applied moments at Support A.

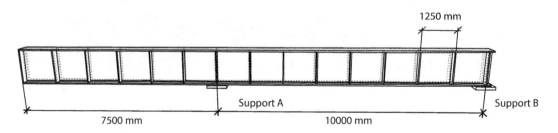

Figure 5.10 Side view of plate girder with cantilever section.

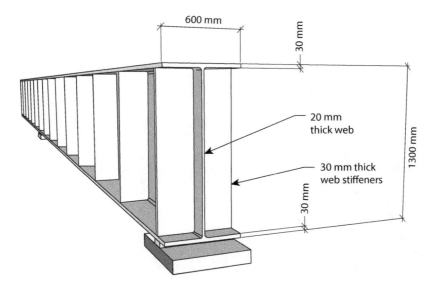

Figure 5.11 End view of plate girder showing cross-section dimensions.

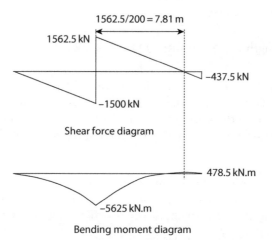

Figure 5.12 Shear force and bending moment diagrams (Example 5.3).

1. First take moments about each support to determine the reactions at Supports A and B labelled in Figure 5.10

$$R_A = \frac{200 \times 17.5^2 \times 0.5}{10} = 3062.5 \text{ kN}$$

$$R_B = \frac{200 \times 17.5 \times 1.25}{10} = 437.5 \text{ kN}$$

Using these reactions, the shear force and bending moment diagrams are calculated, as shown in Figure 5.12.

2. The web stiffeners are simply supported on three edges and free on one; therefore, the buckling coefficient shown in Figure 5.6a applies

$$k = 0.425 + (b/a)^2 = 0.425 + \left(\frac{300 - 10}{1300 - (2 \times 30)} \right)^2 = 0.480$$

The elastic critical buckling stress from Equation 5.3 is

$$\sigma_{cr} = k \times 190000 \times \left(\frac{t}{b} \right)^2 = 0.480 \times 190000 \times \left(\frac{30}{300 - 10} \right)^2 = 976 \text{ N/mm}^2$$

And the design stress from Equation 5.4 is

$$\sigma_{Rd} = \left(\frac{1}{\sigma_y} + \frac{1}{\sigma_{cr}} \right)^{-1} = \left(\frac{1}{265} + \frac{1}{976} \right)^{-1} = 208.4 \text{ N/mm}^2$$

The compression strength is equal to σ_{Rd} multiplied by the stiffener cross-sectional area (30 mm thick and 600 mm wide), i.e.,

$$\text{Compression strength} = 208.4 \times 30 \times 600 \times 10^{-3} = 3751 \text{ kN}$$

The reaction at A is 3062.5 kN, which is less than 3751 kN; therefore, the web stiffener is strong enough. The full width of the stiffener and web was used in the above calculation (600 mm), rather than the stiffener width (580 mm). This reflects the very conservative nature of this calculation, which ignores the contribution of the web to the bearing capacity.

3. The maximum shear force and maximum bending moment both coincide at support A; therefore, the combined effects of shear buckling and buckling due to bending moments must be checked. The first step is to check the resistance against bending stresses.

The second moment of area of the girder is

$$I = \frac{600 \times 1300^3}{12} - \frac{(600 - 20) \times (1300 - 60)^3}{12} = 17.70 \times 10^9 \text{ mm}^4$$

The maximum applied bending stress (σ_{Ed}) in the 1240 mm deep web is

$$\sigma_{Ed} = \frac{M \times y}{I} = \frac{5625 \times 10^6 \times 0.5 \times 1240}{17.70 \times 10^9} = 197 \text{ N/mm}^2$$

The buckling coefficient for pure bending (see Figure 5.6c) is 23.9 and from Equation 5.3

$$\sigma_{cr} = 23.9 \times 190000 \times \left(\frac{20}{1240} \right)^2 = 1181 \text{ N/mm}^2$$

and from Equation 5.4

$$\sigma_{Rd} = \left(\frac{1}{265} + \frac{1}{1181}\right)^{-1} = 216 \text{ N/mm}^2$$

Since $\sigma_{Rd} > \sigma_{Ed}$, the web can resist the applied moments without buckling.

The second step is to check for shear buckling. The applied shear stress (τ_{Ed}) is approximately

$$\tau_{Ed} = \frac{V_{Ed}}{\text{web area}} \tag{5.14}$$

$$\tau_{Ed} = \frac{1562.5 \times 10^3}{1240 \times 20} = 63 \text{ N/mm}^2$$

The web plate length and widths are

$$a = 1250 - 30 = 1220 \text{ mm}$$

$$b = 1300 - 2 \times 30 = 1240 \text{ mm}$$

Since $a \leq b$, Equation 5.8 applies and

$$k = 5.34\left(\frac{1240}{1220}\right)^2 + 4 = 9.52$$

From Equation 5.6

$$\tau_{cr} = 9.52 \times 190000 \times \left(\frac{20}{1240}\right)^2 = 471 \text{ N/mm}^2$$

The yield value of shear stress, from Equation 5.9, is

$$\tau_y = \sigma_y/\sqrt{3} = 153 \text{ N/mm}^2$$

$\tau_{cr} > \tau_y$; therefore, failure will be by yielding and the design stress, $\tau_{Rd} = 153$ N/mm^2.

Since $\tau_{Ed} \ll \tau_{Rd}$, the web is easily strong enough in pure shear, although this ignores effects of bending stresses. The final check is for the combined effects using Equation 5.12

$$\left(\frac{\sigma_{Ed}}{\sigma_{Rd}}\right)^2 + \left(\frac{\tau_{Ed}}{\tau_{Rd}}\right)^2 \leq 1.0$$

$$\left(\frac{197}{216}\right)^2 + \left(\frac{63}{153}\right)^2 = 1.00$$

Therefore, this cross section just passes the combined bending and shear check.

4. The bottom flange also needs checking to determine if it can resist the applied bending stress, which is

$$\sigma_{Ed} = \frac{M \times y}{I} = \frac{5625 \times 10^6 \times 650}{17.7 \times 10^9} = 207 \text{ N/mm}^2$$

The flange can be regarded as simply supported along the boundary with the web. Thus, the buckling coefficient shown in Figure 5.6a should be used and

$$k = 0.425 + \left(\frac{300 - 10}{1250 - 30} \right)^2 = 0.482$$

From Equation 5.3

$$\sigma_{cr} = 0.482 \times 190000 \times \left(\frac{30}{290} \right)^2 = 980 \text{ N/mm}^2$$

The design stress is

$$\sigma_{Rd} = \left(\frac{1}{265} + \frac{1}{980} \right)^{-1} = 209 \text{ N/mm}^2$$

This design strength (209 N/mm²) is slightly higher than the applied stress of 207 N/mm² and the section is just about adequate.

Example 5.4: Box girder

A steel box girder cantilevers 25 m over a support and resists a self-weight of 10 kN/m (unfactored). An imposed load of 8 kN/m (unfactored) is also applied along one outside edge of the girder (1.2 m from the centre line). The girder is stiffened by web stiffeners welded to the outside of the section at 1250 mm centres. The yield stress is 275 N/mm² and the girder is sketched in Figure 5.13.

1. Determine the shear force, bending moment and torsional moment at the base of the cantilever.
2. Determine if the buckling shear stress in the web exceeds the applied shear stress.
3. Determine if the buckling compression strength of the web exceeds the applied compression stress.
4. Determine the adequacy of the web for resisting the combined shear and compression.
5. Determine the adequacy of the beam flange to resist the applied stresses.

1. From Equation 1.2, the ULS uniformly distributed load is

$$w_{uls} = 1.35 \times g_k + 1.5 \times q_k = 1.35 \times 10 + 1.5 \times 8 = 25.5 \text{ kN/m}$$

and the shear force is

$$V_{Ed} = w \times L = 25 \times 25.5 = 637.5 \text{ kN}$$

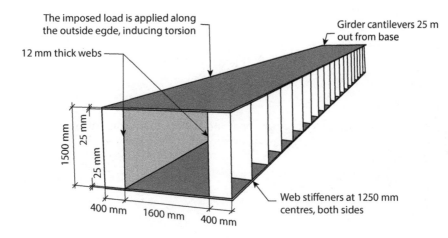

The imposed load is applied along the outside egde, inducing torsion

12 mm thick webs

Girder cantilevers 25 m out from base

1500 mm

25 mm

25 mm

400 mm   1600 mm   400 mm

Web stiffeners at 1250 mm centres, both sides

Figure 5.13 End view of a box girder.

The moment at the base of the cantilever is

$$M_{Ed} = \frac{wL^2}{2} = \frac{25.5 \times 25^2}{2} = 7969 \text{ kN.m}$$

In this example the imposed load is applied along the outside edge inducing a ULS torsional load of

$$w = 1.5 \times q_k = 1.5 \times 8 = 12 \text{ kN/m}$$

and a torsional moment of

$$T = 12 \times 25 \times 1.2 = 360 \text{ kN.m}$$

2. Checking shear buckling of the web: since $a \leq b$, the buckling coefficient from Equation 5.8 is

$$k = 5.34\left(\frac{b}{a}\right)^2 + 4 = 5.34\left(\frac{1450}{1250}\right)^2 + 4 = 11.2$$

From Equation 5.6

$$\tau_{cr} = 11.2 \times 190000 \times \left(\frac{12}{1450}\right)^2 = 146 \text{ N/mm}^2$$

The yield shear stress is

$$\tau_y = \frac{f_y}{\sqrt{3}} = \frac{275}{\sqrt{3}} = 158.7 \text{ N/mm}^2$$

Since $\tau_y > \tau_{cr}$, failure will be by shear buckling rather than yielding. The applied shear stress from Equation 5.14 is approximately

$$\tau_{Ed} = \frac{V_{Ed}}{\text{web area}} = \frac{637.5 \times 10^3}{2 \times 1450 \times 12} = 18.3 \text{ N/mm}^2$$

Shear stress is also developed by the torsional moment. In fact, one of the reasons for using box girders is the high torsional stiffness they provide. The torsional shear stress for a closed (hollow) cross section is

$$\tau = \frac{T}{2tA_m} \qquad (5.15)$$

where
  T is the torsional moment.
  t is the wall thickness at the point where the stress is required.
  A_m is the area enclosed by the midline through the section as shown in Figure 5.14.

A_m should be calculated using the centre line of the plates, i.e.,

$$A_m = 1475 \times 1588 = 2.34 \times 10^6 \text{mm}^2$$

From Equation 5.15, the shear stress due to the torsional moment is

$$\tau = \frac{360 \times 10^6}{2 \times 12 \times 2.34 \times 10^6} = 6.4 \text{ N/mm}^2$$

The combined shear stress is

$$\tau_{Ed} = 18.3 + 6.4 = 24.7 \text{ N/mm}^2$$

Since $\tau_{Ed} \ll 146$ N/mm², this is a clear pass.

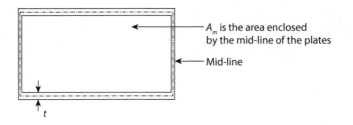

Figure 5.14 Determination of enclosed area for closed cells.

3. The second moment of area of the girder is

$$I = \frac{2400 \times 1500^3}{12} - \frac{(2400 - 2 \times 12) \times (1500 - 2 \times 25)^3}{12} = 71.4 \times 10^9 \text{ mm}^4$$

The applied compression stress in the extreme fibre of the web is

$$\sigma_{Ed} = \frac{M \times y}{I} = \frac{7969 \times 10^6 \times (750 - 25)}{71.4 \times 10^9} = 80.9 \text{ N/mm}^2$$

The elastic critical buckling coefficient, k, is 23.9 (Figure 5.6c) and from Equation 5.3

$$\sigma_{cr} = 23.9 \times 190 \times 10^3 \times \left(\frac{12}{1450}\right)^2 = 311 \text{ N/mm}^2$$

The design stress from Equation 5.4 is

$$\sigma_{Rd} = \left(\frac{1}{275} + \frac{1}{311}\right)^{-1} = 145.9 \text{ N/mm}^2$$

Since the applied bending stress (80.9 N/mm²) is less than the buckling stress (145.9 N/mm²), the section passes this check.

4. The combined shear and bending in the web plate is checked using Equation 5.12, where the design shear stress is the lesser of τ_{cr} (146 N/mm²) and τ_y (158.7 N/mm²); therefore, τ_{Rd} = 146 N/mm² and

$$\left(\frac{\sigma_{Ed}}{\sigma_{Rd}}\right)^2 + \left(\frac{\tau_{Ed}}{\tau_{Rd}}\right)^2 \leq 1.0$$

$$\left(\frac{80.9}{145.9}\right)^2 + \left(\frac{24.7}{146}\right)^2 = 0.33$$

Since this is less than 1.0, this is a pass.

5. The applied compression stress in the flange is

$$\sigma_{Ed} = \frac{M \times y}{I} = \frac{7969 \times 10^6 \times 750}{71.4 \times 10^9} = 83.7 \text{ N/mm}^2$$

The buckling of the outside section of the flange, is checked using the buckling coefficient equation shown in Figure 5.6a

$$k = 0.425 + \left(\frac{400}{1250}\right)^2 = 0.527$$

And from Equation 5.3

$$\sigma_{cr} = 0.527 \times 190 \times 10^3 \times \left(\frac{25}{400}\right)^2 = 391 \text{ N/mm}^2$$

The design stress from Equation 5.4 is

$$\sigma_{Rd} = \left(\frac{1}{275} + \frac{1}{391}\right)^{-1} = 161.5 \text{ N/mm}^2$$

Since this is greater than the applied stress of 83.7 N/mm², this is a pass.

It is also necessary to check buckling of the inside section of the flange, which in this case is governed by Figure 5.6b, where $k = 4.0$. From Equation 5.3

$$\sigma_{cr} = 4.0 \times 190 \times 10^3 \times \left(\frac{25}{1600}\right)^2 = 185 \text{ N/mm}^2$$

From Equation 5.4, the design strength is

$$\sigma_{Rd} = \left(\frac{1}{275} + \frac{1}{185}\right)^{-1} = 110.6 \text{ N/mm}^2$$

Since this greater than the applied stress of 83.7 N/mm², this is also a pass.

Finally, it is necessary from a completeness perspective to check for the combined effects of torsional shear and compression in stresses the flange. From Equation 5.15, the applied torsional shear stress is

$$\tau_{Ed} = \frac{360 \times 10^6}{2 \times 25 \times 2.34 \times 10^6} = 3.1 \text{ N/mm}^2$$

Checking shear buckling of flange caused by the torsional shear stress using Equation 5.7

$$k = 5.34 + 4\left(\frac{b}{a}\right)^2 = 5.34$$

Since the internal section of the flange is not stiffened, b/a tends to zero (1600/25,000) and from Equation 5.6

$$\tau_{cr} = 5.34 \times 190 \times 10^3 \times \left(\frac{25}{1576}\right)^2 = 255 \text{ N/mm}^2$$

The yield shear stress from Equation 5.9 is

$$\tau_y = 275/\sqrt{3} = 159 \text{ N/mm}^2$$

Therefore, this governs. Finally, check the combined compression and shear using Equation 5.13, where the design shear stress is the lesser of τ_{cr} and τ_y; therefore, $\tau_{Rd} = 159$ N/mm²

$$\frac{\sigma_{Ed}}{\sigma_{Rd}} + \left(\frac{\tau_{Ed}}{\tau_{Rd}}\right)^2 \leq 1.0$$

$$\frac{83.7}{110.6} + \left(\frac{3.1}{159}\right)^2 = 0.76$$

Since this is less than 1.0, this is a pass.

5.4 BUCKLING OF STIFFENED PLATES IN COMPRESSION

A stiffened plate is very weak about the unstiffened axis and this causes the dish-shaped buckling mode sketched in Figure 5.15. Due to this low weak axis stiffness, the restraint to buckling provided by the side walls of the box girder has little effect on strength and the buckling capacity can be determined assuming the stiffened plate is equivalent to a column spanning between the internal diaphragms, i.e., ignoring the side walls completely. This will involve a conservatism of the order of 2% in comparison with a full stiffened plate assessment. The corresponding Euler elastic critical buckling stress is

$$\sigma_{cr} = \frac{\pi^2 E I_{st}}{L_{cr}^2 A_{st}} \tag{5.16}$$

where
I_{st} is the second moment of area of the stiffened panel.
A_{st} is the area of the stiffened plate including the area of stiffeners.
L_{cr} is the distance between transverse supports, which are the cross-beams shown in Figure 5.2 or the distance between the internal diaphragms shown in Figure 5.15.

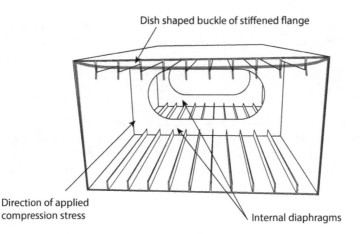

Dish shaped buckle of stiffened flange

Direction of applied compression stress

Internal diaphragms

Figure 5.15 Section through a box girder showing the buckled shape of the top flange when subjected to axial compression.

The Gordon–Rankine equation (Equation 5.4) is then used to estimate the buckling strength of this 'equivalent column'. Whilst this will provide an estimate of the stress at which plate buckling will occur, it will not check for buckling of the individual plate stiffeners or plate material between the stiffeners. Therefore, plate buckling checks also need to be carried out for the individual plates in the panel using the approach described earlier for plate girders.

5.5 BUCKLING OF STIFFENED PLATES IN SHEAR

Box girders are often used due to their high torsional stiffness and as such they can be subject to high shear stresses from torsion, as well as direct shear forces. Since they are thin walled, the stiffened panels may be prone to shear buckling. The elastic critical shear buckling stress is

$$\tau_{cr} = k \frac{\pi^2 E}{12(1-\upsilon^2)} \left(\frac{t}{b}\right)^2$$

which for steel becomes

$$\tau_{cr} = k \times 190000 \times \left(\frac{t}{b}\right)^2 \tag{5.17}$$

The buckling coefficient, k, depends on the aspect ratio of the stiffened plate.

If $a \ge b$
$$k = 5.34 + 4\left(\frac{b}{a}\right)^2 + k_{st} \tag{5.18}$$

If $a \le b$
$$k = 5.34\left(\frac{b}{a}\right)^2 + 4 + k_{st} \tag{5.19}$$

where
  a is the stiffened panel length.
  b is the panel width.
  t is the panel thickness.
  k_{st} is the stiffened panel buckling coefficient.

For the common condition of stiffened plates with three or more longitudinal stiffeners, the buckling coefficient is

$$k_{st} = 9\left(\frac{b}{a}\right)^2 \sqrt[4]{\left(\frac{I_{st}}{t^3 b}\right)^3} \tag{5.20}$$

Imperfections, such as weld shrinkage stresses have much less effect on shear buckling than compression buckling. If the plate is bounded on all four edges by strong plates, as can be the case in box girder bridges, then stiffened plates can possess a significant post-elastic reserve of strength. This means that it is safe to take the design stress (τ_{Rd}) as the lesser of either the elastic critical shear stress or the yield value of shear stress, i.e.,

$$\tau_{Rd} = \tau_{cr} \text{ or } \frac{\sigma_y}{\sqrt{3}} \tag{5.21}$$

This only works if the stiffened plate is bounded on all four edges by strong plates.

5.6 STIFFENED PANELS SUBJECTED TO SHEAR AND COMPRESSION STRESSES

It is important to check the combined effects of shear and compression and an approximate check for combined loading can be performed using the following (empirical) interaction equation

$$\frac{\sigma_{Ed}}{\sigma_{Rd}} + \left(\frac{\tau_{Ed}}{\tau_{Rd}}\right)^2 \leq 1.0 \tag{5.22}$$

where σ_{Ed} and τ_{Ed} are the applied compression and shear stresses, respectively. The design compression stress, σ_{Rd}, should be taken as the whole panel buckling stress determined from Section 5.4 or the outer plate buckling stress for buckling between stiffeners, whichever is lesser, and τ_{Rd} should be the whole panel buckling stress determined from Section 5.5.

Example 5.5: Box girder bridge

Figure 5.16 shows a section through a bridge fabricated from a 70 mm thick plate, which has a yield stress of 325 N/mm². The plates are stiffened with 20 mm thick stiffeners, which have a yield stress of 345 N/mm².

1. Determine the second moment of area of the bottom stiffened plate sketched in Figure 5.17.
2. Determine if this plate can resist a compression stress (applied along the axis of the girder) of 170 N/mm².
3. Determine if this plate can resist a shear stress of 50 N/mm².
4. Can the plate resist the compression and shear stresses from points (1) and (2) combined?

1. To apply the parallel axis theorem to a stiffened plate, it is simpler to consider each plate group separately. The areas (A) and distances (y) between centroids of each element and the bottom of the main plate are

Main plate: $A = 6500 \times 70 = 455000$ mm²

$y = 35$ mm

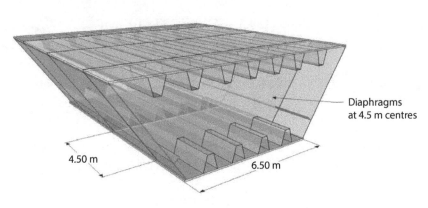

Diaphragms at 4.5 m centres

4.50 m

6.50 m

Figure 5.16 Cross section through a box girder.

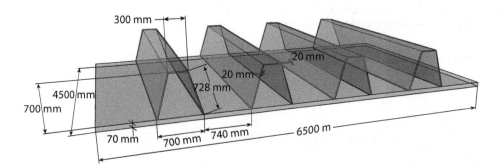

Figure 5.17 The stiffened bottom flange of the box girder shown in Figure 5.16.

Stiffener webs: $A = 8 \times 20 \times 728 = 116480$ mm²

$$y = 70 + 700 / 2 = 420 \text{ mm}$$

Top plates: $A = 4 \times 20 \times 300 = 24000$ mm²

$$y = 700 + 70 - 20 / 2 = 760 \text{ mm}$$

Total area: $A_{st} = 595.48 \times 10^3$ mm²

Taking moments about the bottom of the main plate, the distance between the bottom of the main plate and the centroid of area of the stiffened plate is

$$\overline{y} \times 595.48 \times 10^3 = 455000 \times 35 + 116480 \times 420 + 24000 \times 760$$

$$\overline{y} = 139.5 \text{ mm}$$

Now apply the parallel axis theorem to determine the second moment of area of the stiffened plate, I_{st}

$$I_{plate} = \frac{6500 \times 70^3}{12} + 455000 \times (139.5 - 35)^2 = 5.155 \times 10^9 \text{ mm}^4$$

$$I_{stiffener, web} = 8 \times \left[\frac{20 \times 700^3}{12} + 728 \times 20 \times (350 + 70 - 139.5)^2 \right] = 13.738 \times 10^9 \text{ mm}^4$$

$$I_{stiffener, top} = 4 \times \left[\frac{300 \times 20^3}{12} + 20 \times 300 \times (760 - 139.5)^2 \right] = 9.241 \times 10^9 \text{ mm}^4$$

$$I_{st} = I_{plate} + I_{stiffener, web} + I_{stiffener, top} = 28.134 \times 10^9 \text{ mm}^4$$

2. The calculation of the buckling stresses is split into five steps for simplicity.

Step 1: Check buckling of the whole stiffened plate between diaphragms. The elastic critical buckling stress of the 'equivalent column' from Equation 5.16 is

$$\sigma_{cr} = \frac{\pi^2 E I_{st}}{L_{cr}^2 A_{st}}$$

$$\sigma_{cr} = \frac{\pi^2 \times 210000 \times 28.134 \times 10^9}{4500^2 \times 595.48 \times 10^3} = 4835 \text{ N/mm}^2$$

Note that in the above calculation, the effective length (L_{cr}) was the distance between the diaphragms (see Figure 5.16). This is because the equivalent column effectively spans these points. Using the Gordon–Rankine approximation (Equation 5.4), the design stress is

$$\sigma_{Rd} = \left(\frac{1}{325} + \frac{1}{4835} \right)^{-1} = 305 \text{ N/mm}^2$$

Since the applied stress $\sigma_{Ed} \ll \sigma_{Rd}$ (i.e. 170 N/mm² ≪ 305 N/mm²), there is little danger of the plate buckling in this mode.

Step 2: Buckling of the main plate between the stiffeners. The bottom flange of the box girder is assumed as being pinned at the boundaries with the stiffeners and the diaphragms; therefore the critical buckling coefficient (k) is taken as equal to 4.0 throughout (see Figure 5.6b). The bottom plate is 70 mm thick (t) and the maximum distance between stiffeners (b) is 740 mm. From Equation 5.2

$$\sigma_{cr} = 4.0 \times 190 \times 10^3 \times \left(\frac{70}{740} \right)^2 = 6801 \text{ N/mm}^2$$

and the design stress from Equation 5.4 is

$$\sigma_{Rd} = \left(\frac{1}{325} + \frac{1}{6801} \right)^{-1} = 310 \text{ N/mm}^2$$

Step 3: Buckling of the stiffener webs. The stiffener webs are bounded on all four edges, three of which are simply supported while one is effectively built in. One edge is considered as built in because the main plate is 70 mm thick, which is much more than the 20 mm thick stiffener. The main plate can therefore be regarded as a rigid boundary from a buckling perspective and $k = 5.41$ (see Figure 5.6f). The webs are 20 mm thick and 728 mm wide; therefore, from Equation 5.2

$$\sigma_{cr} = 5.41 \times 190 \times 10^3 \times \left(\frac{20}{728} \right)^2 = 776 \text{ N/mm}^2$$

And the design stress from Equation 5.4 is

$$\sigma_{Rd} = \left(\frac{1}{345} + \frac{1}{776}\right)^{-1} = 239 \text{ N/mm}^2$$

Step 4: Buckling of the stiffener flanges. The stiffener flanges are bounded on all four edges by simple supports; therefore, $k = 4.0$ and from Equations 5.2 and 5.4

$$\sigma_{cr} = 4.0 \times 190 \times 10^3 \times \left(\frac{20}{300}\right)^2 = 3378 \text{ N/mm}^2$$

$$\sigma_{Rd} = \left(\frac{1}{345} + \frac{1}{3378}\right)^{-1} = 313 \text{ N/mm}^2$$

Step 5: Identify the critical buckling stress. The analysis presented in the above steps indicates that buckling is likely to occur at the following stresses:

Whole panel → 305 N/mm²

Main plate between the stiffeners → 310 N/mm²

Stiffener web → 239 N/mm²

Stiffener flange → 313 N/mm²

The stiffener webs are critical because they buckle at the lowest stress. However, the stiffened plate is capable of resisting an applied compression stress of 170 N/mm², as is asked in the question. It can be seen that this design does fall short of a fully stressed design and it would be more efficient if the stiffener web were adjusted to increase the buckling stress. This could be done by slightly reducing the length of the stiffener webs.

3. Considering shear buckling of the bottom flange of the box girder, from Equation 5.20 the stiffened plate buckling coefficient is

$$k_{st} = 9\left(\frac{b}{a}\right)^2 \sqrt[4]{\left(\frac{I_{st}}{t^3 b}\right)^3} = 9 \times \left(\frac{6500}{4500}\right)^2 \sqrt[4]{\left(\frac{28.134 \times 10^9}{70^3 \times 6500}\right)^3} = 125.7$$

Since $a \leq b$, from Equation 5.19

$$k = 5.34\left(\frac{b}{a}\right)^2 + 4 + k_{st} = 5.34\left(\frac{6.5}{4.5}\right)^2 + 4 + 125.7 = 141$$

and the elastic critical shear stress from Equation 5.17 is

$$\tau_{cr} = k \times 190000 \times \left(\frac{t}{b}\right)^2 = 141 \times 190000 \times \left(\frac{70}{6500}\right)^2 = 3107 \text{ N/mm}^2$$

The yield shear stress from Equation 5.9 is

$$\tau_y = \frac{\sigma_y}{\sqrt{3}} = \frac{325}{\sqrt{3}} = 188 \text{ N/mm}^2$$

From Equation 5.21, the design shear stress is the lesser of τ_{cr} and τ_y, i.e., $\tau_{Rd} = 188$ N/mm². Since this is greater than the applied shear stress (τ_{Ed}) of 50 N/mm², the plate is easily passes this check.

4. The interaction equation for combined compression and shear, Equation 5.22, is

$$\frac{\sigma_{Ed}}{\sigma_{Rd}} + \left(\frac{\tau_{Ed}}{\tau_{Rd}}\right)^2 \leq 1.0$$

The applied shear stresses will be highest in the outer plate; therefore, the lesser of the following compression buckling stresses is used:

Whole panel → 305 N/mm²

Main plate between the stiffeners → 310 N/mm²

Therefore, Equation 5.22 becomes

$$\frac{170}{305} + \left(\frac{50}{188}\right)^2 = 0.63$$

Since this is less than 1.0, this simple analysis would indicate that the plate should able to resist these combined stresses; however, these calculations do not include any allowance for bending moments resulting from loads supported directly by the stiffened panel. This problem is considered next.

5.7 STIFFENED PLATES WITH LATERAL LOADS

Longitudinal stiffeners run along the length of box girders and these are very flexible in bending about their weak axis. For this reason, stiffened plates will effectively span between the cross frames when supporting lateral loads (see Figure 5.2 and Figure 5.15). It is necessary to check the stability of these plates subjected to compression as well as bending. An approximate check can be achieved by treating the stiffened plate as an equivalent 'beam column' spanning between cross frames or diaphragms. This simplification ignores the restraint provided by the sides of the box girder, although this involves only a minor degree of conservatism. The process is spilt into three separate steps:

Step 1: Determine the buckling stress for the stiffened plate, σ_{Rd}. This would be the stress required to cause overall buckling of the stiffened panel, or the stress required to cause buckling of one of the individual plate elements, whichever is lowest. This process was illustrated in Example 5.5(2).

Step 2: Determine the moment capacity of the stiffened panel. To carry out a 'beam-column analysis', the moment capacity of the stiffened panel must be established. The stiffened panel is a thin-walled structure; therefore, the moment capacity will be lower than the elastic moment capacity, because local buckling rather than yielding will initiate failure. For this reason, the buckling stress of the stiffeners must be determined. The moment capacity is estimated by setting the extreme fibre stress equal to the design buckling stress for the stiffener (i.e. not σ_y).

Step 3: Check the stability of the stiffened plate when subjected to combined bending and compression. This check is essentially the same as that used for the beam-columns as described in Section 3.2. The axial force resisted by the stiffened plate (σ_{Ed}) will amplify any deflection resulting from the applied moments (M_{Ed}) that occur due to the lateral load. A small (secondary) bending moment will be developed due to the axial force multiplied by this bending-induced deflection. This secondary moment can be included using the same amplification factor used in many different buckling situations in engineering. The amplification factor is

$$\alpha = \frac{1}{1 - \dfrac{\sigma_{Ed}}{\sigma_{cr}}} \tag{5.23}$$

This can also included in design expressions as

$$\alpha = \frac{\sigma_{cr}}{\sigma_{cr} - \sigma_{Ed}}$$

where
   σ_{Ed} is the applied compression stress.
   σ_{cr} is the elastic critical buckling stress for the stiffened panel, from Equation 5.16.

The following interaction equation can be used to assess the ability of the stiffened plate to resist combined loading

$$\frac{\sigma_{Ed}}{\sigma_{Rd}} + \frac{\alpha M_{Ed}}{M_{Rd}} \leq 1 \tag{5.24}$$

The design strength σ_{Rd}, should be taken as the lowest value of compression buckling stress calculated for the stiffened panel, as illustrated in Example 5.5(2).

5.8 STIFFENED PANELS IN SHEAR, COMPRESSION AND BENDING

It is important to check that cross sections will not buckle when subjected to the combined effects of shear stresses, compression stresses and/or bending stresses. An approximate check can be obtained by combining Equations 5.22 and 5.24, i.e.,

$$\frac{\sigma_{Ed}}{\sigma_{Rd}} + \frac{\alpha M_{Ed}}{M_{Rd}} + \left(\frac{\tau_{Ed}}{\tau_{Rd}}\right)^2 \leq 1.0 \tag{5.25}$$

The design strength, σ_{Rd}, should be taken as the lowest value of compression buckling stress calculated for the stiffened panel and τ_{Rd} should be the whole panel buckling stress.

Example 5.6: Box girder bridge

A box girder comprises top and bottom stiffened plates with diaphragms spaced at 4000 mm centres (see Figure 5.18). The yield stress is 355 N/mm². For the top and bottom stiffened plates shown in Figure 5.18, determine

1. The second moment of area.
2. The axial compression stress that results in buckling.
3. The shear stress that would result in shear failure.
4. Determine if the stiffened plate can resist 70 N/mm² of shear stress combined with 180 N/mm² of applied compression stress.
5. The moment capacity of the stiffened plate.
6. Estimate what uniformly distributed load the top plate can support in addition to resisting the combined loading from point (4) above.

1. The areas (A) and distances from plate centroids to the bottom of the panel (y) are as follows (Figure 5.19):

Main plate: $A = 3900 \times 15 = 58500$ mm²

$y = 7.5$ mm

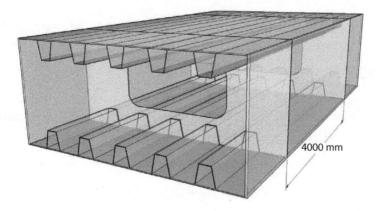

4000 mm

Figure 5.18 Perspective view showing a section through a box girder.

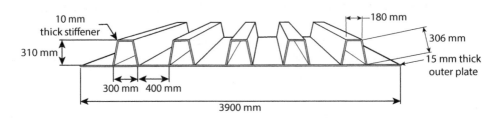

Figure 5.19 Dimensions of the stiffened flanges (top and bottom) of the girder sketched in Figure 5.18.

Stiffener webs: $A = 10 \times 10 \times 306 = 30600$ mm²

$y = 15 + 310 / 2 = 170$ mm

Top plates: $A = 5 \times 10 \times 180 = 9000$ mm²

$y = 15 + 305 = 320$ mm

Total area of top plate including stiffeners: $A_{st} = 98100$ mm²

Taking moments about the bottom of the main plate to locate the centroid of the stiffened plate:

$$\bar{y} = \frac{58500 \times 7.5 + 30600 \times 165 + 9000 \times 320}{98100} = 85.3 \text{ mm}$$

Applying the parallel axis theorem to determine the second moment of area of the stiffened plate, I_{st}:

$$I_{\text{plate}} = \frac{3900 \times 15^3}{12} + 58500 \times (85.3 - 7.5)^2 = 355 \times 10^6 \text{ mm}^4$$

$$I_{\text{stiffener, web}} = 10 \times \left[\frac{10 \times 310^3}{12} + 306 \times 10 \times (15 + 155 - 85.3)^2 \right] = 468 \times 10^6 \text{ mm}^4$$

$$I_{\text{stiffener, top}} = 5 \times \left[\frac{180 \times 10^3}{12} + 10 \times 180 \times (15 + 305 - 85.3)^2 \right] = 496 \times 10^6 \text{ mm}^4$$

$$I_{st} = I_{\text{plate}} + I_{\text{stiffener, web}} + I_{\text{stiffener, top}} = 1319 \times 10^6 \text{ mm}^4$$

2. The calculation of the buckling stress of the stiffened plate is broken up into five steps for simplicity.

Step 1: Check buckling of the whole stiffened plate between diaphragms.

The elastic critical buckling stress of the equivalent column from Equation 5.16 is

$$\sigma_{cr} = \frac{\pi^2 E I_{st}}{L_{cr}^2 A_{st}} = \frac{\pi^2 \times 210000 \times 1319 \times 10^6}{4000^2 \times 98100} = 1742 \text{ N/mm}^2$$

and the design stress from Equation 5.4 is

$$\sigma_{Rd} = \left(\frac{1}{355} + \frac{1}{1742} \right)^{-1} = 295 \text{ N/mm}^2$$

Step 2: Check buckling of the main plate between the stiffeners.

The critical buckling coefficient, k, is taken as equal to 4.0 (see Figure 5.6b). The bottom plate is 15 mm thick and the maximum distance between stiffeners is 400 mm; therefore, from Equation 5.2

$$\sigma_{cr} = 4.0 \times 190 \times 10^3 \times \left(\frac{15}{400} \right)^2 = 1069 \text{ N/mm}^2$$

and the design buckling stress (Equation 5.4)

$$\sigma_{Rd} = \left(\frac{1}{355} + \frac{1}{1069} \right)^{-1} = 266 \text{ N/mm}^2$$

Step 3: Check buckling of the stiffener webs.

All four of the stiffener web boundaries are regarded as pinned; therefore, $k = 4.0$ (see Figure 5.6b). The webs are 10 mm thick and 306 mm wide; therefore,

$$\sigma_{cr} = 4.0 \times 190 \times 10^3 \times \left(\frac{10}{306} \right)^2 = 812 \text{ N/mm}^2$$

The design stress is

$$\sigma_{Rd} = \left(\frac{1}{355} + \frac{1}{812} \right)^{-1} = 247 \text{ N/mm}^2$$

Step 4: Check buckling of stiffener flanges.

The stiffener flanges are bounded on all four edges by simple supports; therefore, $k = 4.0$ (see Figure 5.6b) and

$$\sigma_{cr} = 4.0 \times 190 \times 10^3 \times \left(\frac{10}{180} \right)^2 = 2346 \text{ N/mm}^2$$

$$\sigma_{Rd} = \left(\frac{1}{355} + \frac{1}{2346} \right)^{-1} = 308 \text{ N/mm}^2$$

Step 5: Identify the critical buckling stress.

The analysis showed that the buckling stresses are

Whole panel $\rightarrow$ 295 N/mm²

Main plate between the stiffeners $\rightarrow$ 266 N/mm²

Stiffener web → 247 N/mm²

Stiffener flange → 308 N/mm²

Of these, the stiffener webs are critical and the maximum compression stress that can be resisted by the top and bottom panels of the box girder is approximately 247 N/mm².

3. Considering shear buckling from Equation 5.20

$$k_{st} = 9 \times \left(\frac{3900}{4000}\right)^2 \sqrt[4]{\left(\frac{1319 \times 10^6}{15^3 \times 3900}\right)^3} = 271$$

Since $a \geq b$, Equation 5.18 is used to determine the shear buckling coefficient

$$k = 5.34 + 4\left(\frac{b}{a}\right)^2 + k_{st} = 5.34 + 4\left(\frac{3900}{4000}\right)^2 + 271 = 280$$

From Equation 5.17, the elastic critical shear stress is

$$\tau_{cr} = 280 \times 190000 \times \left(\frac{15}{3900}\right)^2 = 787 \text{ N/mm}^2$$

And from Equation 5.9, the yield shear stress is

$$\tau_y = \frac{\sigma_y}{\sqrt{3}} = \frac{355}{\sqrt{3}} = 205 \text{ N/mm}^2$$

Finally from Equation 5.21, the design shear stress is the lesser of τ_{cr} and τ_y; therefore, $\tau_{Rd} = 205$ N/mm².

4. The interaction equation for combined compression and shear, Equation 5.22, is

$$\frac{\sigma_{Ed}}{\sigma_{Rd}} + \left(\frac{\tau_{Ed}}{\tau_{Rd}}\right)^2 \leq 1.0$$

In this check the lesser of the compression buckling stresses for the whole panel (295 N/mm²) or the main plate between the stiffeners (266 N/mm²) is used, i.e.,

$$\frac{180}{266} + \left(\frac{70}{205}\right)^2 = 0.79$$

Since this is less than 1.0, the stiffened plate should be approximately strong enough.

5. The maximum bending stress will occur furthest from the neutral axis, which in this case is the top of the stiffener flanges. These are predicted to buckle at a stress of 308 N/mm² and so the moment capacity is

$$M_{Rd} = \frac{\sigma \times I}{y} = \frac{308 \times 1319 \times 10^6}{310 + 15 - 85.3} \times 10^{-6} = 1695 \text{ kN.m}$$

6. The interaction equation, Equation 5.25, must be populated to determine the maximum value for the applied moment, M_{Ed}

$$\frac{\sigma_{Ed}}{\sigma_{Rd}} + \frac{\alpha_y M_{Ed}}{M_{Rd}} + \left(\frac{\tau_{Ed}}{\tau_{Rd}}\right)^2 \leq 1$$

The moment amplification factor (α) is calculated using the applied compression stress and the elastic critical stress calculated for the whole stiffened plate, i.e., from Equation 5.23

$$\alpha = \frac{1}{1 - \dfrac{\sigma_{Ed}}{\sigma_{cr}}} = \frac{1}{1 - \dfrac{180}{1742}} = 1.12$$

In this case, the web is resisting the bending stresses, and since the web has a low buckling stress we will use it in the combined stress check to be safe, i.e.,

$$\frac{180}{247} + \frac{1.12 \times M_{Ed}}{1695} + \left(\frac{70}{205}\right)^2 \leq 1.0$$

$$M_{Ed} \leq 234 \text{ kN.m}$$

This moment is dependent on the loading configuration. For a UDL, the critical hogging moment occurs over the diaphragms and it can conservatively be assumed to be equal to

$$M_{Ed} \approx \frac{wL^2}{8}$$

$$\therefore \quad w = \frac{M_{Ed} \times 8}{L^2} = \frac{234 \times 8}{4.0^2} = 117 \text{ kN/m}$$

or

$$w = \frac{117}{3.9} = 30.0 \text{ kN/m}^2$$

This calculation is probably conservative, although it does provide a quick approximation of the lateral load capacity, which in this case is approximately 3 tonnes per m² (1 tonne is 9.81 kN).

Example 5.7: Box girder

Figure 5.20 shows a box girder. The top and bottom flanges are 3500 mm wide and are stiffened by 10 longitudinal stiffeners, which are placed at 318 mm centres. Each stiffener is 150 mm deep and 15 mm thick. Diaphragms spaced at 3000 mm centres also stiffen the box girder. The yield stress is 275 N/mm². Considering the bottom-stiffened plate shown in Figure 5.20:

1. Determine the second moment of area of the stiffened plate.
2. Determine the compression stress required to cause this plate to buckle between cross frames.
3. Determine the compression stress required to cause the outer plate to buckle between longitudinal stiffeners.
4. Determine the compression stress required to cause the longitudinal plate stiffeners to buckle.
5. Determine the stress required to cause shear buckling of the stiffened plate.
6. The box girder is subject to a torsional moment of 1700 kN.m, in addition to a horizontal shear of 4000 kN. Determine the maximum applied shear stress in the stiffened plate under this applied loading.
7. Determine the maximum shear stress that the stiffened plate can resist in combination with an axial compression stress of 125 N/mm².
8. Determine the elastic moment capacity of the stiffened flange.
9. Determine the maximum UDL that can be supported by the stiffened panel, in addition to a compression stress of 125 N/mm².
10. Recalculate point (9) above but include the loading from point (6), i.e., consider combined bending, compression and shear

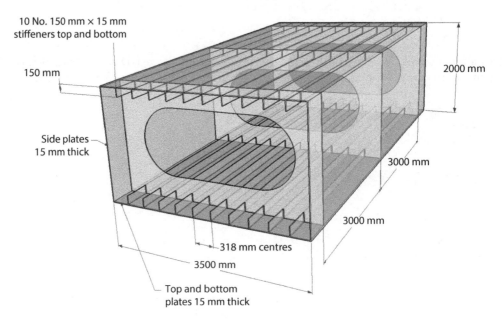

Figure 5.20 Perspective view showing a section through a box girder.

1. The cross-sectional area of the stiffened plate is

$$A_{st} = 3500 \times 15 + 150 \times 15 \times 10 = 75000 \text{ mm}^2$$

The distance to the centroid of the stiffened plate from the outer face of the main plate is

$$\bar{y} = \frac{15 \times 3500 \times 7.5 + 10 \times 15 \times 150 \times (15 + 75)}{75000} = 32.25 \text{ mm}$$

Using the parallel axis theorem, the second moment of area of the stiffened plate is

$$I_{st} = 15 \times 3500 \times (32.25 - 7.5)^2 + 10 \times 15 \times 150 \times (15 + 75 - 32.25)^2$$

$$+ \frac{3500 \times 15^3}{12} + \frac{10 \times 15 \times 150^3}{12} = 150 \times 10^6 \text{ mm}^4$$

2. Considering buckling of the whole plate between the diaphragms, the elastic critical buckling stress from Equation 5.16 is

$$\sigma_{cr} = \frac{\pi^2 \times 210000 \times 150 \times 10^6}{3000^2 \times 75000} = 460.6 \text{ N/mm}^2$$

The design stress from Equation 5.4 is

$$\sigma_{Rd} = \left(\frac{1}{275} + \frac{1}{460.6} \right)^{-1} = 172 \text{ N/mm}^2$$

3. For buckling of the main plate between stiffeners, the coefficient shown in Figure 5.6b applies; therefore,

$$\sigma_{cr} = 4.0 \times 190000 \left(\frac{15}{318} \right)^2 = 1691 \text{ N/mm}^2$$

and

$$\sigma_{Rd} = \left(\frac{1}{275} + \frac{1}{1691} \right)^{-1} = 237 \text{ N/mm}^2$$

4. Consider buckling of the stiffeners. The outer plate is of the same thickness of the stiffeners (both 15 mm); therefore, it is assumed that the boundary between the plates is simply supported, instead of fixed, and the coefficient shown in Figure 5.6a applies, i.e.,

$$k = 0.425 + (b/a)^2 = 0.425 + (150/3000)^2 = 0.428$$

If the outer plate had been significantly thicker than the stiffeners, then coefficient in Figure 5.6e could have been used. From Equations 5.2 and 5.4

$$\sigma_{cr} = 0.428 \times 190000 \left(\frac{15}{150} \right)^2 = 813 \text{ N/mm}^2$$

$$\sigma_{Rd} = \left(\frac{1}{275} + \frac{1}{813} \right)^{-1} = 205.5 \text{ N/mm}^2$$

Cautionary note. This buckling stress of 205.5 N/mm² may be an overestimate of strength because of a form of buckling known as 'local torsional buckling' (see Figure 5.21) (Horne, 1977). In this mode, stiffeners interact and buckle at a lower stress than predicted by this simple treatment of buckling. Local torsional buckling is too complex for this book, although the easiest way to overcome it is to use closed (box-shaped) stiffeners, such as those used in the bridge shown in Figure 5.22. Closed stiffeners (Figure 5.19) do not suffer from local torsional buckling and this may in part explain their popularity.

5. From Equation 5.20

$$k_{st} = 9 \times \left(\frac{3500}{3000} \right)^2 \sqrt[4]{\left(\frac{150 \times 10^6}{15^3 \times 3500} \right)^3} = 82.4$$

Since $a \leq b$, the shear buckling coefficient is determined using Equation 5.19

$$k = 5.34 \left(\frac{3500}{3000} \right)^2 + 4 + 82.4 = 93.7$$

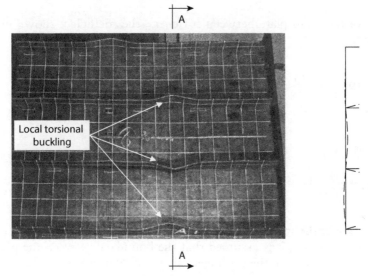

Stiffened plate under uniform compression   Section A–A

Figure 5.21 Local torsional buckling of plate stiffeners.

Figure 5.22 This bridge over the River Severn in the United Kingdom uses box-shaped stiffeners, similar to those shown in Figure 5.18.

The elastic critical shear stress from Equation 5.17 is

$$\tau_{cr} = 93.7 \times 190000 \times \left(\frac{15}{3500}\right)^2 = 327 \text{ N/mm}^2$$

The yield shear stress from Equation 5.9 is

$$\tau_y = \frac{275}{\sqrt{3}} = 158.8 \text{ N/mm}^2$$

The design shear stress is the lesser of τ_{cr} and τ_y; therefore, $\tau_{Rd} = 158.8$ N/mm². This shows that shear stresses will cause yielding before shear buckling of the stiffened panel.

6. The shear stress due to the horizontal shear force is approximately equal to the shear force divided by the cross-sectional areas of the top and bottom plates, i.e.,

$$\tau = \frac{4000 \times 10^3}{2 \times 3500 \times 15} = 38.1 \text{ N/mm}^2$$

It should be noted that if the full shear stress equation (Equation 6.11) is used to calculate the maximum shear stress in the middle of the plate, the stress rises to 45.3 N/mm². The shear stress due to torsion also needs to be calculated. The area enclosed by the box girder is

$$A_m = 3500 \times 2000 = 7 \times 10^6 \text{ mm}^2$$

and the stress due to the torsional moment from Equation 5.15 is

$$\tau = \frac{T}{2tA_m} = \frac{1700 \times 10^6}{2 \times 15 \times 7 \times 10^6} = 8.1 \text{ N/mm}^2$$

The total shear stress is

$$\tau_{Ed} = 38.1 + 8.1 = 46.2 \text{ N/mm}^2$$

7. The above analysis showed that the buckling stresses are

Whole panel → 172 N/mm²

Main plate between the stiffeners → 237 N/mm²

Plate stiffeners → 205.5 N/mm²

Of these, the whole panel buckling is critical; therefore, σ_{Rd} = 172 N/mm² and from Equation 5.22

$$\frac{\sigma_{Ed}}{\sigma_{Rd}} + \left(\frac{\tau_{Ed}}{\tau_{Rd}}\right)^2 \leq 1.0$$

$$\frac{125}{172} + \left(\frac{\tau_{Ed}}{158.8}\right)^2 = 1.0$$

$$\tau_{Ed} = 83.0 \text{ N/mm}^2$$

Since the applied shear stress determined above was 46.2 N/mm², this plate should be more than adequate for this load combination.

8. The moment capacity of the stiffened plate will be limited by buckling of the stiffeners, since they experience the highest bending stresses, being furthest from the neutral axis, as well as having the lowest buckling stress. The buckling stress of the stiffeners was calculated as equal to 205.5 N/mm²; therefore, the moment capacity of the stiffened plate is

$$M_{Rd} = \frac{\sigma I}{y} = \frac{205.5 \times 150 \times 10^6}{15 + 150 - 32.25} \times 10^{-6} = 232 \text{ kN.m}$$

9. Equation 5.24 needs to be completed to determine the limiting value for the applied moment, M_{Ed}

$$\frac{\sigma_{Ed}}{\sigma_{Rd}} + \frac{\alpha_y M_{Ed}}{M_{Rd}} \leq 1$$

The moment amplification factor (α) is calculated using the applied compression stress and the elastic critical stress calculated for the whole stiffened plate in part (2) of this question, i.e., from Equation 5.23

$$\alpha = \frac{1}{1 - \dfrac{\sigma_{Ed}}{\sigma_{cr}}} = \frac{1}{1 - \dfrac{125}{460.6}} = 1.372$$

The applied bending stress is $\sigma_{Ed} = 125$ N/mm^2 and the compression strength is $\sigma_{Rd} = 172$ N/mm^2; therefore, Equation 5.24 becomes

$$\frac{125}{172} + \frac{1.372 \times M_{Ed}}{232} \leq 1$$

This solves to $M_{Ed} \leq 46.2$ kN.m. This is dependent on the loading configuration and can conservatively be assumed as

$$M_{Ed} = \frac{wL^2}{8}$$

$$w = \frac{M_{Ed} \times 8}{L^2} = \frac{46.2 \times 8}{3.0^2} = 41.1 \text{ kN/m}$$

or by dividing by the width, the UDL is

$$w = \frac{41.1}{3.5} = 11.7 \text{ kN/m}^2$$

10. The basic interaction, Equation 5.25, is

$$\frac{\sigma_{Ed}}{\sigma_{Rd}} + \frac{\alpha_y M_{Ed}}{M_{Rd}} + \left(\frac{\tau_{Ed}}{\tau_{Rd}}\right)^2 \leq 1.0$$

This becomes

$$\frac{125}{172} + \frac{1.372 \times M_{Ed}}{232} + \left(\frac{46.2}{158.8}\right)^2 \leq 1.0$$

This rearranges to provide a maximum applied moment for the stiffened panel, $M_{Ed} \leq 31.9$ kN.m, and

$$w = \frac{31.9 \times 8}{3.0^2} = 28.4 \text{ kN/m}$$

or dividing by the width, the UDL is

$$w = \frac{28.4}{3.5} = 8.1 \text{ kN/m}^2$$

This shows that addition of shear stresses has caused the load capacity to fall from 11.7 kN/m^2 to 8.1 kN/m^2.

Problems

Solutions to these problems are provided at https://www.crcpress.com/9781498741217

P.5.1. A 34 m long I-section steel plate girder spans between two supports (see Figure 5.23). The girder is restrained along its length against lateral torsional buckling and is subjected to a uniformly distributed load (inclusive of self-weight and load factors) of 180 kN/m. The yield stress is 440 N/mm².

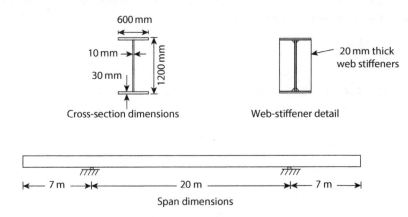

Figure 5.23 Bridge question.

a. Determine the shear force and moment at midspan and at the supports.
b. 20 mm thick web stiffeners are located at 3 m centres in the centre of the span; determine if the girder flanges can resist the compression force due to bending without buckling.
c. Determine if the girder web can resist the applied loading at midspan without buckling.
d. Web stiffeners are located at 0.75 m centres in the region of the supports; determine if the girder web can resist the applied loading at the supports without buckling.
e. If the web is stiffened with 20 mm plates that extend across the entire width of the beam (600 mm; see web stiffener detail in Figure 5.2), determine if the stiffeners are sufficiently strong to resist the applied loads.

Ans. (a) 4590 kN.m, 4410 kN.m, 1260 kN, 1800 kN, (b) 290.5 > 203 N/mm² ∴ pass, (c) 194.6 > 192 N/mm² pass, (d) $(184.8/194.6)^2 + (158/249)^2 = 1.30 > 1.0$ fail and (e) 217.5 < 255 N/mm² fail.

P.5.2. Figure 5.24 shows a box girder. The top and bottom flanges are 4000 mm wide and are stiffened by 13 stiffeners that are 150 mm deep, 20 mm thick and located at 300 mm centres. Diaphragms spaced at 4000 mm centres stiffen the girder. The yield stress is 265 N/mm². Considering the bottom-stiffened plate shown in Figure 5.25:
a. Determine the second moment of area of the stiffened plate.
b. Determine the compression stress required to cause the stiffened plate to buckle between diaphragms.
c. Determine the compression stress required to cause the outer plate to buckle between stiffeners.

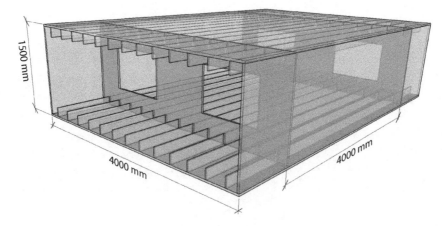

Figure 5.24 Part of a long box girder showing internal diaphragms spaced at 4000 mm centres.

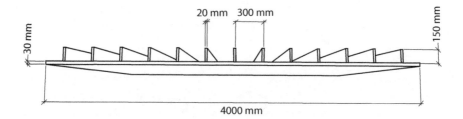

Figure 5.25 Bottom-stiffened plate from girder shown in Figure 5.24.

d. Determine the stress to cause the stiffeners to buckle (hint: Figure 5.6e).
e. Determine the elastic critical shear buckling stress of the whole panel.
f. Determine the yield value of shear stress.
g. Determine the shear stress that the stiffened plates can resist in addition to a compressive stress of 100 N/mm².

Ans. (a) 320.5×10^6 mm⁴, (b) 131.5 N/mm², (c) 256 N/mm², (d) 249 N/mm², (e) 316.4 N/mm², (f) 153 N/mm² and (g) 75 N/mm².

REFERENCES

Allen, H. G. and Bulson, P. S., 1980. *Background to Buckling.* United States: McGraw Hill.
Horne, M. R., 1977. *Ciria Guide 3: Structural Action in Box Girders.* London: Construction Industry Research and Information Association.

Chapter 6

Composite structures

Figure 6.1a shows a steel I-beam supporting a reinforced concrete slab. When the I-beam flexes, the top flange goes into compression and shortens, whereas the bottom fibre of the slab goes into tension and lengthens. This causes slippage to occur between the beam and slab. The strength and stiffness can be increased by preventing this slippage, as illustrated in Figure 6.1b. In this type of 'composite beam', the I-beam and slab act together to resist moments. Slippage is prevented by casting the concrete around metal fixings welded to the top flange of the I-beam. These take the form of 'shear studs', which are metal rods with flat heads welded to the beam before the concrete is cast.

Composite beams can be either propped or unpropped during casting of the slab. If unpropped, then wet concrete is supported by the steel section alone. Contractors prefer this, because propping costs money; however, unpropped construction is more complicated to design. A lack of understanding by engineers has occasionally led to beams being over-stressed. This chapter will attempt to explain the theory of how to design unpropped beams safely.

6.1 EFFECTIVE WIDTH

When a composite beam flexes, the slab acts like a compression flange. In order to determine bending strength, it is necessary to establish the width of slab that contributes to

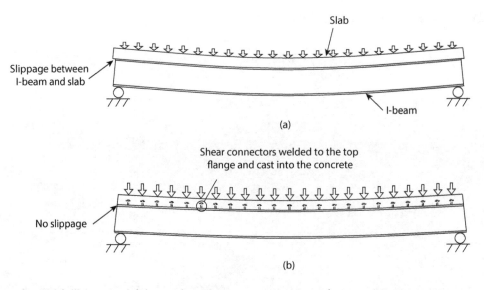

Figure 6.1 Illustration of the steel concrete non-composite and composite arrangements. (a) Conventional beam (without shear studs) and (b) composite beam.

the flexural strength. This is known as the *effective width* and is illustrated in Figure 6.2. The effective width should be taken as the lesser of either:

1. One-eighth of the *effective span* on each side; see Figure 6.2a, or as illustrated in Figure 6.2b for edge beams
2. Half the distance between beam centre lines

The *effective span* is the distance between points of zero bending moments. This is the distance between supports for simply supported beams. For beams that are continuous over supports, such as bridge beams, the effective span is taken as the distance between points of contraflexure (zero moments).

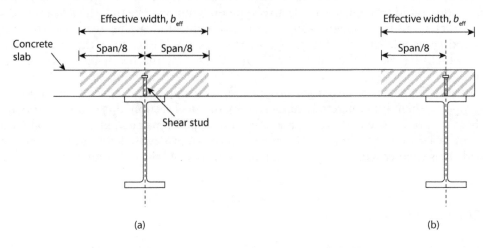

Figure 6.2 Section through composite beams to illustrate the concept of *effective width*. (a) Internal beam and (b) edge beam.

Example 6.1: Calculation of effective width

Figure 6.3 shows a framing arrangement for the floor of a building. The steel beams are designed to act compositely with the slab and the beam connections are 'nominally pinned'. Determine the effective widths of the beams labelled A to F.

In this example the beam connections are nominally pinned; therefore, the effective span is simply the beam span.

Beam A: This is an edge beam similar to that shown in Figure 6.2b. On the south side, the slab extends 0.75 m from the beam centre line; therefore, the effective width on the south side is 0.75 m. On the north side, the effective width is the lesser of either half the spacing between Beams A and B (2.0 m) or span/8 (1.5 m). Therefore, b_{eff} = 0.75 + 1.5 = 2.25 m.

Beam B: On the north side, Beams B and C are spaced at 2.5 m centres; therefore, the maximum effective width = 2.5/2 = 1.25 m. The total effective width = 1.25 + 1.5 = 2.75 m.

Beam C: b_{eff} = 1.25 + 1.25 = 2.5 m

Beam D: b_{eff} = 0.75 + 1.25 = 2.0 m

Beam E: b_{eff} = 2 × span/8 = 1.25 m

Beam F: b_{eff} = 2 × span/8 = 1.0 m

6.2 SERVICEABILITY LIMIT STATE DESIGN

During serviceability limit state (SLS) design, loads are unfactored and the primary objectives are twofold:

- Ensure that the beams remains elastic
- Prevent excessive deflection

The first step is to calculate the second moment of area of the composite beam, I_{comp}. To do this, the *method of transformed sections* is used, whereby the effective width of the slab is converted to an equivalent area of steel. This conversion is made by multiplying the effective width (b_{eff}) by the modular ratio, which is the ratio between Young's modulus for concrete

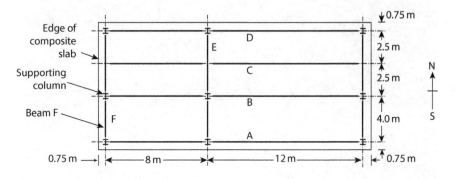

Figure 6.3 Plan view of a floor showing beam layout and composite slab.

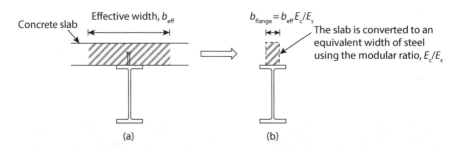

Figure 6.4 Using the method of transformed sections to calculate the width of the top flange. (a) Cross section through a composite floor beam and (b) the transformed section.

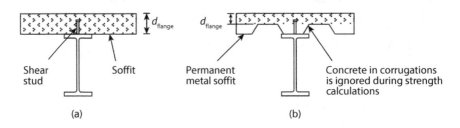

Figure 6.5 Two main types of composite beam. (a) Flat soffit and (b) profiled metal deck soffit.

and steel, E_c/E_s (see Figure 6.4). Once the cross section is converted into a fully 'steel' section, it is necessary to take moments of area about the top of the slab to determine the position of the centroid of area. The *parallel axis theorem* is then used to calculate the second moment of area of the composite section.

Figure 6.5a shows a composite beam with a solid concrete slab, although it is more common in buildings to cast concrete slabs onto corrugated metal sheeting, which is left in place after concreting. This is known as *profiled metal decking* and it is widely used in the construction of buildings; see Figure 6.5b and Figure 6.6. In this book, the concrete within the corrugations is ignored during analysis in order to speed up calculations.

Figure 6.6 Profiled metal deck composite slab under construction (shear studs indicate beam positions).

Composite beams are vulnerable to *lateral torsional buckling* (LTB) when supporting wet concrete. Beams with the profiled metal decking spanning along the direction of span (Figure 6.5b) are laterally unrestrained when supporting wet concrete. If the decking corrugations are perpendicular to the span, then the profiled metal deck will provide some restraint against buckling. If in doubt, then all beams should be considered as laterally unrestrained under wet concrete loads. Although LTB is important, it is not considered again in this chapter because it is dealt with in Chapter 2.

Example 6.2: SLS calculations for a beam in a building

A floor slab is supported by beams spaced at 3 m centres and spanning 6 m between simple supports. The beams support the full weight of the 200 mm deep slab, in addition to an imposed load of 5 kN/m². The beams and the slab are joined together by shear studs to act compositely and the beams are unpropped when the concrete slab is cast.

1. Determine the beam stresses and mid-span deflection when supporting wet concrete.
2. Determine the position of the neutral axis and the second moment of area of the composite beam.
3. Sketch the elastic bending stress distribution under unfactored dead and imposed loads (i.e. SLS loads).
4. Determine the mid-span deflection under SLS loads.

Basic data
Second moment of area of the steel beam = 15,000 cm⁴
Cross-sectional area of steel section = 6000 mm²
Beam self-weight = 0.5 kN/m
Depth of the steel section = 400 mm
Young's modulus for steel, E_s = 210,000 N/mm²
Young's modulus for the concrete, E_c = 25,000 N/mm²

1. The load (w) and midspan moment (M) under the wet weight of the concrete are

$$w = 0.5 + 25 \times 0.2 \times 3.0 = 15.5 \text{ kN/m (15.5 N/mm)}$$

$$M = \frac{15.5 \times 6^2}{8} = 69.75 \text{ kN.m}$$

Using the engineer's beam equation, the maximum bending stresses (σ) in the steel section under the wet concrete are

$$\sigma = \frac{M \times (\pm) \dfrac{d}{2}}{I} \tag{6.1}$$

$$\sigma = \frac{69.75 \times 10^6 \times (\pm 200)}{15000 \times 10^4} = \pm 93.0 \text{ N/mm}^2$$

and the midspan deflection from Equation 1.6 is

$$\delta = \frac{5wL^4}{384EI}$$

$$\delta = \frac{5 \times 15.5 \times 6000^4}{384 \times 210000 \times 15000 \times 10^4} = 8.3 \text{ mm}$$

2. The effective width of the slab is the lesser of either the beam spacing (3 m) or

$$b_{\text{eff}} = 2 \times \frac{L}{8} = \frac{2 \times 6000}{8} = 1500 \text{ mm}$$

Using the *modular ratio* required by the *method of transformed sections*, the width of the top flange of the transformed section (Figure 6.4b) is

$$b_{\text{flange}} = b_{\text{eff}} \times \frac{E_c}{E_s} \tag{6.2}$$

$$b_{\text{flange}} = 1500 \times \frac{25000}{210000} = 178.6 \text{ mm}$$

The neutral axis depth labelled x in Figure 6.7b is located by taking moments of area about a reference point, which in this case is the top of the slab. For simplicity, it is assumed that the slab remains uncracked (i.e. that it takes a tensile load) and taking moments of area about the top of the slab:

$$x \times (178.6 \times 200 + 6000) = 178.6 \times 200 \times \frac{200}{2} + 6000\left(200 + \frac{400}{2}\right)$$

$$x = 143 \text{ mm}$$

The second moment of area is now calculated using the *parallel axis theorem*, i.e.,

$$I_{\text{comp}} = 178.6 \times 200 \times \left(143 - \frac{200}{2}\right)^2 + \frac{178.6 \times 200^3}{12} + 6000 \times \left(200 + \frac{400}{2} - 143\right)^2$$

$$+ 15000 \times 10^4$$

$$I_{\text{comp}} = 731.4 \times 10^6 \text{ mm}^4$$

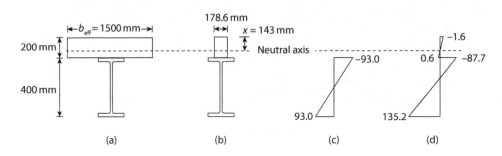

Figure 6.7 Calculation of elastic bending stresses. (a) The composite beam, (b) the transformed section, (c) wet concrete stress and (d) SLS stress distribution.

3. Since the beams are spaced at 3 m centres, the 5 kN/m² imposed load develops the following beam load:

$$w = 3 \times 5 = 15 \text{ kN/m}$$

and the corresponding midspan moment is

$$M_{Ed} = \frac{15 \times 6^2}{8} = 67.5 \text{ kN.m}$$

The stress at the top of the slab is determined using the engineer's beam equation and by reapplying the modular ratio to convert the transformed section stresses back to concrete stresses, i.e.,

$$\sigma = \frac{E_c}{E_s} \times \frac{M_y}{I} \tag{6.3}$$

$$\sigma = \frac{25000}{210000} \times \frac{67.5 \times 10^6 \times (-)143}{731.4 \times 10^6} = -1.6 \text{ N/mm}^2$$

And the stress in the concrete at the bottom of the slab is

$$\sigma = \frac{25000}{210000} \times \frac{67.5 \times 10^6 \times (200 - 143)}{731.4 \times 10^6} = 0.6 \text{ N/mm}^2$$

The stress at the top of the steel section includes the stress developed when supporting the 'wet' concrete (−93 N/mm²) and the imposed load stress

$$\sigma = -93.0 + \frac{67.5 \times 10^6 \times (200 - 143)}{731.4 \times 10^6} = -87.7 \text{ N/mm}^2$$

At the bottom of the section,

$$\sigma = +93.0 + \frac{67.5 \times 10^6 \times (200 + 400 - 143)}{731.4 \times 10^6} = +135.2 \text{ N/mm}^2$$

The stresses under the wet concrete and under full SLS loads are shown graphically in Figure 6.7. Since the maximum stress (135.2 N/mm²) is much less than the yield stress (275 N/mm²), the beam will remain elastic under SLS loads.

4. The total deflection = wet concrete deflection (8.3 mm) + imposed load deflection. Remember to use different values of second moment of area in these calculations, with I_{steel} used for wet concrete and I_{comp} used for imposed load deflection, i.e.,

$$\delta = 8.3 + \frac{5 \times 15 \times 6000^4}{384 \times 210000 \times 731.4 \times 10^6} = 9.9 \text{ mm}$$

6.3 ULS BENDING STRENGTH

During ULS design, it is necessary to check the strength and stability under wet concrete (if unpropped), as well as when loaded by the fully factored dead and imposed loads. There are two approaches to the calculation of the ULS moment capacity of a composite beam. The first is elastic design and this is mainly used for bridges. The second is to utilise the plastic bending strength, but this is only used for buildings.

The mode of construction affects the stress distribution, i.e., if the section is propped or unpropped during construction. However, the absence of propping during construction is generally assumed to have little effect on the plastic moment capacity. Specifically, the moment capacity of an unpropped beam can be the same as that of a propped beam, providing the steel section is Class 1 or Class 2 and providing certain other conditions are met. In order to determine the moment capacity of a steel beam, it is necessary to classify the cross section as described in Section 2.2. If the cross section is Class 1 or Class 2, the full plastic moment capacity of the composite section can be utilised. If Class 3, the maximum stress must be limited to the yield stress, and if Class 4 the maximum stress is defined using the methods described in Chapter 5 for thin-walled structures.

Plastic moment capacity. The plastic moment capacity is easy to calculate. The first step is to determine the tensile strength of the steel section, *T*. The partial safety factor for steel is normally set at 1.0; therefore,

$$T = f_y A_s \tag{6.4}$$

where
A_s is the area of the steel section.
f_y is the yield stress.

The next step is to determine the depth to the *plastic neutral axis*, shown as *x* in Figure 6.8. Remember that this is on a different position to the *elastic neutral axis* (see Figure 6.7b).

If design is based on concrete cylinder strengths, as is the case for the Eurocodes, then the design crushing stress

$$f_{cd} = \frac{0.85 f_{ck}}{\gamma_c}$$

Since the partial safety factor for concrete (γ_c) is 1.5

$$f_{cd} = 0.567 f_{ck} \tag{6.5}$$

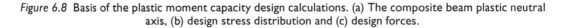

Figure 6.8 Basis of the plastic moment capacity design calculations. (a) The composite beam plastic neutral axis, (b) design stress distribution and (c) design forces.

Assuming that the plastic neutral axis lies within the slab, then the compression force in the concrete is

$$C = 0.567 f_{ck} \times b_{eff} \times x \qquad (6.6)$$

where b_{eff} is the effective width (see Section 6.1). If no external axial force is applied, then from horizontal equilibrium

$$C = T \qquad (6.7)$$

Combining Equations 6.6 and 6.7, the depth to the plastic neutral axis is

$$x = \frac{T}{0.567 f_{ck} b_{eff}} \qquad (6.8)$$

At this stage, it is necessary to check that x lies within the top part of the slab, i.e., that $x < (d_c - d_p)$; see Figure 6.8. If this is not the case, then Equation 6.6 will need reconfiguring, although this is rarely a problem. The lever arm (z) between T and C (see Figure 6.8c) is

$$z = \frac{d_p}{2} + d_c - \frac{x}{2} \qquad (6.9)$$

where
  d_c is the depth of the concrete slab.
  d_p is the depth of the profiled metal decking; see Figure 6.8.

Finally, the plastic moment capacity is the force multiplied by the lever-arm, i.e.,

$$M_{pl,\,Rd} = T \times z \qquad (6.10)$$

Example 6.3: Calculating the bending strength of a floor beam

A floor slab is supported by (Class 1) I-section beams spaced 3 m apart and spanning 12 m (see Figure 6.9). The composite slab is constructed using a profiled metal decking system similar to that shown in Figure 6.6, with a depth of the profiled metal decking (d_p) of 100 mm. Determine if the beams can support an imposed load of 5.0 kN/m² at the ULS. The beams are unpropped.

Basic data: f_y = 355 N/mm², f_{ck} = 40 N/mm², cross-sectional area of the I-beams = 58.6 cm², weight of steel beam = 0.45 kN/m and unfactored weight of the slab is 3.75 kN/m².

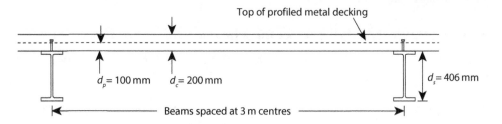

Figure 6.9 Cross section through a composite slab showing beams and metal decking.

From Equation 1.3, the ULS load is

$$w_{uls} = 1.35 \times 0.45 + 3.0 \times (1.35 \times 3.75 + 1.5 \times 5.0) = 38.3 \text{ kN/m}$$

and the corresponding midspan moment is

$$M_{Ed} = \frac{wL^2}{8} = \frac{38.3 \times 12^2}{8} = 689 \text{ kN.m}$$

The effective width of the slab is the lesser of either the beam spacing (3 m) or

$$b_{eff} = 2 \times \frac{L}{8} = \frac{2 \times 12}{8} = 3 \text{ m}$$

The composite moment capacity remains unaffected by the lack of propping, because the beams are Class 1 and are sufficiently ductile to form the plastic stress distribution shown in Figure 6.8b. From Equation 6.4, the tensile strength of the steel section is

$$T = A_s f_y = 5860 \times 355 \times 10^{-3} = 2080 \text{ kN}$$

The neutral axis depth from Equation 6.8 is

$$x = \frac{T}{0.567 f_{ck} b_{eff}}$$

$$x = \frac{2080 \times 10^3}{0.567 \times 40 \times 3000} = 30.6 \text{ mm}$$

Since x is less than 100 mm, the neutral axis is located well within the top half of the slab. The lever arm from Equation 6.9 is

$$z = \frac{d_s}{2} + d_c - \frac{x}{2}$$

$$z = \frac{406}{2} + 200 - \frac{30.6}{2} = 387.7 \text{ mm}$$

From Equation 6.10

$$M_{pl.Rd} = T \times z = 2080 \times 387.7 \times 10^{-3} = 806 \text{ kN.m}$$

Since the applied moment (689 kN.m) is less than the moment capacity (806 kN.m), the beam has sufficient bending strength.

Example 6.4: ULS strength checks for a bridge

Figure 6.10 shows a composite bridge deck that is unpropped during construction and spans 36 m between simple supports. Considering Girder A:

1. Determine if it remains elastic under ULS loads from wet concrete + girder self-weight (2.75 kN/m) + workers and equipment (0.75 kN/m²).
2. Determine the composite beam second moment of area.

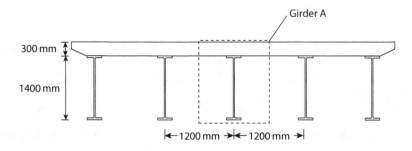

Figure 6.10 Composite bridge beam.

3. Sketch the stress distribution under ULS dead loads + imposed load of 10.5 kN/m². Check to see if the beam remains elastic.

Basic data: Concrete Young's modulus = 25,000 N/mm² and density = 25 kN/m³. Girder A properties: yield stress = 265 N/mm², Young's modulus = 210,000 N/mm², $I = 13,400 \times 10^6$ mm⁴, cross-sectional area = 39,840 mm².

1. From Equation 1.3 the ULS dead and imposed load at the construction stage is

$$w = 1.35 \times 2.75 + 1.35 \times 0.3 \times 1.2 \times 25 + 1.5 \times 0.75 \times 1.2 = 17.2 \text{ kN/m}$$

and the midspan moment is

$$M_{Ed} = \frac{17.2 \times 36^2}{8} = 2786 \text{ kN.m}$$

From Equation 6.1, the maximum stresses in the I-section are

$$\sigma = \frac{2786 \times 10^6 \times (\pm)700}{13400 \times 10^6} = \pm 145.5 \text{ N/mm}^2$$

This is less than the yield stress of 265 N/mm², so the beam will remain elastic. It should therefore be safe, providing the compression flanges are restrained against sideways movement in order to prevent lateral torsional buckling.

2. The effective width of the concrete flange is the lesser of the beam spacing (1200 mm) or 2 × span/8 (7.5 m). The corresponding width of the top flange of the transformed section (see Figure 6.4b) is

$$b_{flange} = \frac{b_{eff} E_c}{E_s} = \frac{1200 \times 25000}{210000} = 143 \text{ mm}$$

The distance from the top of the slab to the elastic neutral axis (x) is determined by taking moments of area about the top of the slab, i.e.,

$$143 \times 300 \times \frac{300}{2} + 39840 \times \left(300 + \frac{1400}{2} \right) = x \times (143 \times 300 + 39840)$$

$x = 559$ mm

The second moment of area is calculated using the *parallel axis theorem*, i.e.,

$$I_{comp} = 143 \times 300 \times \left(559 - \frac{300}{2}\right)^2 + \frac{143 \times 300^3}{12} + 39840 \times \left(300 + \frac{1400}{2} - 559\right)^2 + 13400 \times 10^6$$

$$I_{comp} = 28646 \times 10^6 \, mm^2$$

3. The ULS dead load, when resisting the self-weight of 2.75 kN/m + the wet weight of the slab, is

$$w = 1.35 \times 2.75 + 1.35 \times 0.3 \times 1.2 \times 25 = 15.9 \, kN/m$$

And the corresponding moment is

$$M_{Ed} = \frac{15.9 \times 36^2}{8} = 2576 \, kN.m$$

From Equation 6.1, the stresses in the top and bottom fibres of the section are

$$\sigma = \frac{2576 \times 10^6 \times (\pm)700}{13400 \times 10^6} = \pm 135 \, N/mm^2$$

Note that the second moment of area of only the steel section was used in the above equation and not I_{comp}. The ULS imposed load is

$$w = 1.5 \times 10.5 \times 1.2 = 18.9 \, kN/m$$

and the corresponding moment is

$$M_{Ed} = \frac{18.9 \times 36^2}{8} = 3062 \, kN.m$$

From Equation 6.3, this produces the following stress in the top of the slab

$$\sigma = \frac{E_c}{E_s} \times \frac{M_y}{I_{comp}}$$

$$\sigma = \frac{25000}{210000} \times \frac{3062 \times 10^6 \times (-)559}{28646 \times 10^6} = -7.1 \, N/mm^2$$

The stress in the bottom of the slab is

$$\sigma = \frac{25000}{210000} \times \frac{3062 \times 10^6 \times (-559 + 300)}{28646 \times 10^6} = -3.3 \, N/mm^2$$

At the top of the steel beam,

$$\sigma = -135 + \frac{3062 \times 10^6 \times (-559 + 300)}{28646 \times 10^6} = -162.7 \, N/mm^2$$

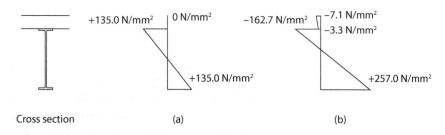

+135.0 N/mm²   0 N/mm²   −162.7 N/mm²   −7.1 N/mm²
                                           −3.3 N/mm²

+135.0 N/mm²   +257.0 N/mm²

Cross section   (a)   (b)

Figure 6.11 ULS stresses in an unpropped composite bridge. (a) Wet concrete stresses and (b) full ULS stresses.

And the bottom of the section

$$\sigma = +135 + \frac{3062 \times 10^6 \times (-559 + 300 + 1400)}{28646 \times 10^6} = +257.0 \text{ N/mm}^2$$

Figure 6.11 shows the stresses when loaded by the ULS loads. The elastic limit is reached when this stress > 265 N/mm²; therefore, the steel should just remain elastic under the ULS loading.

6.4 SHEAR STUD DESIGN

As discussed earlier, slippage between the steel and slab is controlled using shear studs welded to the top flange and cast into the concrete. There are two approaches to the design of these studs:

1. *Elastic design*, in which each stud is designed to resist the shear force without significant deformation. In this approach, the distribution of the shear connectors reflects the shear force distribution in the beam. Figure 6.12a shows the shear force distribution for

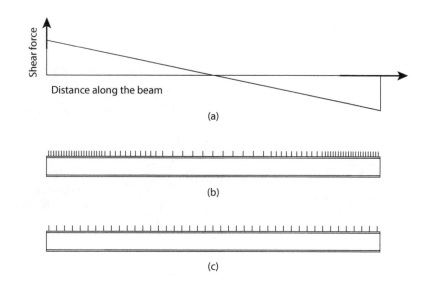

Figure 6.12 Comparison of the different approaches to designing shear studs. (a) Shear force diagram, (b) beam with elastically designed shear studs (stepped spacing to mirror shear force diagram) and (c) beam with plastically designed shear studs (linear spacing).

a simply supported beam subjected to a UDL. Figure 6.12b shows the arrangement of shear studs that are designed elastically, with the density of shear studs increased at the ends to reflect the increased shear force. This approach is used mainly in bridges.

2. *Plastic design*, whereby the shear connectors are uniformly distributed (see Figure 6.12c). In this approach, shear studs need to be ductile in order to distribute the shear force evenly along the beam. This is achieved by using headed shear studs manufactured from ductile steel. Plastic design of shear studs must not be used for bridges because of fatigue. It is also not suitable for beams supporting concentrated loads near supports or moving loads. Despite these limitations it is used for most steel framed multistorey buildings.

6.4.1 Elastic design of the shear studs

The elastic shear stress equation gives the shear stress (τ) a distance y from the neutral axis

$$\tau = \frac{VA'\bar{y}}{b_o I} \tag{6.11}$$

where
  I is the second moment of area (in this case for the composite section).
  V is the shear force.
  b_o is the width of the section at distance y from the neutral axis.
  A' is the area of the section above the distance y from the neutral axis.

Figure 6.13 shows how to apply this equation to the problem of determining the shear stress at the interface between the steel and concrete. In this case, $\bar{y}$ is the distance between the centroid of the slab to the neutral axis of the composite section. From a design perspective, shear flow, which is the shear force per length (τb_o), is more useful; thus from Equation 6.11,

$$\text{Shear flow} = \frac{VA'\bar{y}}{I} \tag{6.12}$$

This can be used to determine the stud spacing, because

$$\text{Shear flow} = \frac{\text{stud strength}}{\text{stud spacing}} \tag{6.13}$$

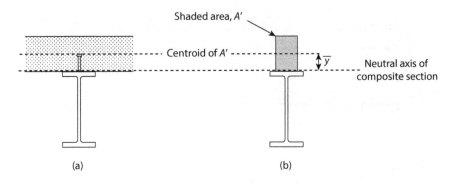

Figure 6.13 Notation used in Equation 6.12. (a) The composite beam and (b) the transformed section.

Example 6.5: Shear stud design for a floor beam

The composite beam shown in Figure 6.14 is constructed using shear connectors with a shear strength of 70 kN spaced at 100 mm intervals. Determine the maximum shear force that the beam can resist before the shear connectors reach their limit.

From Equation 6.13, the shear flow from 70 kN shear studs spaced at 100 mm centres is

$$\text{Shear flow} = \frac{\text{shear stud strength}}{\text{shear stud spacing}} = \frac{70 \times 10^3}{100} = 700 \text{ N/mm}$$

Rearranging Equation 6.12

$$V = \frac{\text{shear flow} \times I}{A'\bar{y}}$$

With reference to Figure 6.13,

$$A' = 366 \times 200 = 73200 \text{ mm}^2$$

$$\bar{y} = 142 - 200/2 = 42 \text{ mm}$$

And the shear studs will become overloaded at a shear of

$$V = \frac{700 \times 615 \times 10^6}{73200 \times 42} \times 10^{-3} = 140 \text{ kN}$$

Example 6.6: Shear stud design with point loads

The composite beam shown in Figure 6.14 supports two concentrated loads in addition to a UDL (see Figure 6.15). Draw the ULS shear force diagram and determine the shear stud spacing at the ends of the beam and in the central low shear region. The beam is *propped* during construction and the shear stud capacity is 100 kN per stud.

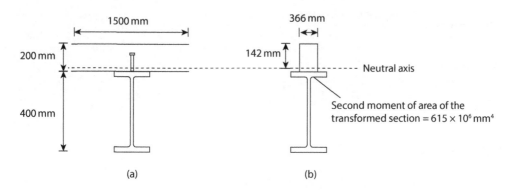

Figure 6.14 Design information. (a) The composite beam and (b) the transformed section.

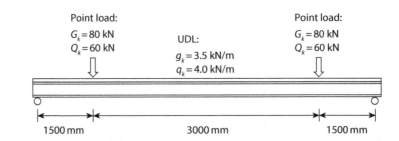

Figure 6.15 Beam with point loads and a uniformly distributed load.

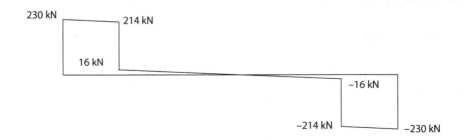

Figure 6.16 Shear force diagram for beam shown in Figure 6.15.

Since the beam has point loads near the supports, elastic design should be used. The beam is propped; therefore, composite action resists dead and imposed loading. Therefore, all the loads are used to determine the shear stud spacing. The ULS point load (P) is

$$P = 1.35G_k + 1.5Q_k = 1.35 \times 80 + 1.5 \times 60 = 198 \text{ kN}$$

And the ULS UDL is

$$w = 1.35g_k + 1.5q_k = 1.35 \times 3.5 + 1.5 \times 4 = 10.7 \text{ kN/m}$$

These loads are used to plot the shear force diagram shown in Figure 6.16.

From Equation 6.12, the shear flow at the supports is

$$\text{Shear flow} = \frac{230 \times 10^3 \times 366 \times 200 \times (142 - 200/2)}{615 \times 10^6} = 1150 \text{ N/mm}$$

From Equation 6.13

$$\text{Shear stud spacing} = \frac{100 \times 10^3}{1150} = 87 \text{ mm}$$

In the central section,

$$\text{Shear flow} = \frac{16 \times 10^3 \times 366 \times 200 \times (142 - 200/2)}{615 \times 10^6} = 80 \text{ N/mm}$$

And the maximum shear stud spacing is

$$\text{Spacing} = \frac{100 \times 10^3}{80} = 1250 \text{ mm}$$

The stud spacing should not be greater than six times the slab depth or 800 mm, in order to prevent the slab separating from the beam; therefore, the maximum spacing in the central region is 800 mm.

6.4.2 Plastic design of shear studs

Shear studs in buildings are designed to maintain strength after a considerable amount of plastic deformation, which means that all the studs can be assumed to develop the full shear force, even if they are evenly spread along the length of a beam, as illustrated in Figure 6.12c. This works with beams supporting uniformly distributed loads or point loads located at the centre of the span, although problems occur in beams supporting point loads away from the centre. Therefore, care needs to be applied when designing studs plastically. Elastic design of studs should be used for bridges, because fatigue is a design issue and therefore stress is tightly controlled.

To develop the required bending strength, the force from shear studs must be equal to the lesser of

1. tensile strength of the steel section
2. compression strength of the concrete flange

In buildings, the tensile strength of the steel section is almost always less than the compressive strength of the concrete flange; therefore, the design of the shear studs is easy.

Example 6.7: Plastic design of shear studs for a uniformly distributed load

A simply supported beam spans 10 m and supports a uniformly distributed load. During the design, the plastic neutral axis was found to be in the slab and the tensile strength of the steel section = 2080 kN. Determine the spacing of the shear studs if 85 kN studs are used.

The total shear force required from the studs is the lesser of either the tensile strength of the section or the compression strength of the slab. In this case, the neutral axis is in the slab; therefore, the tensile strength of the section governs. This means that there must be enough studs (in each half of the beam) to develop 2080 kN of force. Therefore, the number of studs in each half of the beam is

$$\text{Number of studs} = \frac{2080 \text{ kN}}{85 \text{ kN}} = 25$$

The span is 10,000 mm; therefore,

$$\text{Stud spacing} = \frac{10000/2}{25} = 200 \text{ mm}$$

Example 6.8: Plastic design of shear studs with point loads

Figure 6.17a shows the beam layout for a building. During the plastic analysis of Beam A, the neutral axis was found to be in the slab and the tensile strength of the steel section was found to be 2080 kN. Determine the spacing of the shear studs in Beam A if 85 kN studs are used.

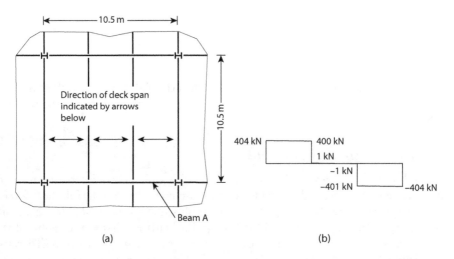

Figure 6.17 Composite beam design for a building. (a) Plan view of floor and (b) shear force diagram for Beam A.

The neutral axis is in the slab. This means that there must be enough shear studs to develop the tensile force in the steel of 2080 kN. Therefore,

$$\text{Number of studs} = \frac{2080 \text{ kN}}{85 \text{ kN}} = 25$$

Figure 6.17b shows the shear force diagram. It reveals that there is very little shear in the central third span. Therefore, the 25 shear studs need to be distributed over the end third spans instead of the half span, which was the case in the previous example. Therefore,

$$\text{Spacing} = \frac{10500/3}{25} = 140 \text{ mm}$$

Shear studs should be spaced at 140 mm centres in the end third spans. The shear force in the central third span is low and studs can be spaced at the maximum allowable spacing.

Example 6.9: Elastic design with a point load

Figure 6.18 shows a slab supported by I-beams spaced at 750 mm centres and spanning 10 m between simple supports. Each beam supports an imposed load of 15 kN positioned at midspan + a uniformly distributed imposed load of 20 kN/m (all unfactored).

1. Calculate the second moment of area of the composite beam.
2. Determine the maximum midspan deflection under SLS dead and imposed loading if the beams are unpropped during construction.
3. Sketch the elastic stress distribution under ULS dead and imposed loads.
4. Determine the maximum shear stud spacing at the supports using elastic theory if the shear stud capacity is 50 kN/stud.
5. Determine the distance from the supports that shear stud spacing can be increased to 300 mm.

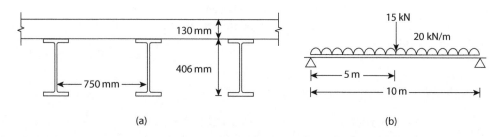

Figure 6.18 Unpropped beam. (a) Cross section through composite beams and (b) imposed loading applied to each beam.

Basic data
Beam self-weight = 0.52 kN/m
Second moment of area of the steel section = 21,508 cm⁴
Cross-sectional area of the steel = 7600 mm²
Young's modulus for the steel = 210,000 N/mm²
Young's modulus for the concrete = 14,000 N/mm²

1. The effective width of the composite beam is the lesser of either the beam spacing (750 mm) or twice span/8 (2500 mm). The width of the transformed section top flange from Equation 6.2 is

$$b_{flange} = 750 \times \frac{14000}{210000} = 50 \text{ mm}$$

The distance (x) between the top of the slab and the centroid of the transformed section is defined by taking moments of area about the top of the slab, i.e.,

$$50 \times 130 \times \frac{130}{2} + 7600 \times (130 + 406/2) = x \times (50 \times 130 + 7600)$$

$$x = 209.5 \text{ mm}$$

Using the *parallel axis theorem*, the second moment of area is

$$I_{comp} = \frac{50 \times 130^3}{12} + 50 \times 130 \times \left(209.5 - \frac{130}{2}\right)^2 + 21508 \times 10^4 + 7600 \times (130 + 203 - 209.5)^2$$

$$I_{comp} = 475.9 \times 10^6 \text{ mm}^4$$

2. The dead load from the slab and beam self-weight is

$$w_{dl} = 25.0 \times 0.75 \times 0.13 + 0.52 = 2.96 \text{ kN/m}$$

Since the beam is unpropped, the deflection is calculated using I for the plain steel section; thus, the midspan deflection is

$$\Delta_{dl} = \frac{5 \times 2.96 \times 10000^4}{384 \times 210000 \times 21508 \times 10^4} = 8.5 \text{ mm}$$

The imposed loads shown in Figure 6.18b are resisted by the composite action between the steel and concrete. Therefore, the composite beam second moment of area is used and the imposed load deflection is

$$\Delta = \frac{5wL^4}{384EI} + \frac{PL^3}{48EI}$$

$$\Delta_{il} = \frac{5 \times 20 \times 10000^4}{384 \times 210000 \times 475.9 \times 10^6} + \frac{15 \times 10^3 \times 10000^3}{48 \times 210000 \times 475.9 \times 10^6} = 29.2 \text{ mm}$$

And the total deflection from the dead and imposed loads is

$$\Delta = 8.5 + 29.2 = 37.7 \text{ mm}$$

3. The ULS dead load from the slab and beam self-weight is

$$w_{dl} = 1.35 \times (25.0 \times 0.75 \times 0.13 + 0.52) = 3.99 \text{ kN/m}$$

And the dead load midspan moment is

$$M_{dl} = \frac{3.99 \times 10^2}{8} = 49.9 \text{ kN.m}$$

The beam supports the dead loads without assistance from the concrete slab; therefore, the stresses in the outer fibres are calculated using the second moment of area of the steel section only, from Equation 6.1

$$\sigma = \frac{49.9 \times 10^6 \times (\pm 203)}{21508 \times 10^4} = \pm 47.1 \text{ N/mm}^2$$

The midspan moment due to the factored imposed load is

$$M = \frac{wL^2}{8} + \frac{PL}{4}$$

$$M_{il} = \frac{1.5 \times 20 \times 10^2}{8} + \frac{1.5 \times 15 \times 10}{4} = 431.25 \text{ kN.m}$$

Stress at the top of the slab is

$$\sigma = \frac{E_c}{E_s} \times \frac{M_y}{I}$$

$$\sigma = \frac{14000}{210000} \times \frac{431.25 \times 10^6 \times (-)209.5}{475.9 \times 10^6} = -12.7 \text{ N/mm}^2$$

At the bottom of the slab,

$$\sigma = \frac{14000}{210000} \times \frac{431.25 \times 10^6 \times (-209.5 + 130)}{475.9 \times 10^6} = -4.80 \text{ N/mm}^2$$

Top of the steel

$$\sigma = -47.1 + \frac{431.25 \times 10^6 \times (-209.5 + 130)}{475.9 \times 10^6} = -119.1 \, \text{N/mm}^2$$

Bottom of the steel

$$\sigma = +47.1 + \frac{431.25 \times 10^6 \times (-209.5 + 130 + 406)}{475.9 \times 10^6} = 343.0 \, \text{N/mm}^2$$

These stresses are shown graphically in Figure 6.19. The most common grade steel (S355) has a yield stress of 355 N/mm²; therefore, the beam should remain elastic if this grade is used.

4. The shear studs develop no stress when the beam supports wet concrete. Therefore, *only* loads applied *after* the concrete *has hardened* contribute to shear stud loading. The ULS shear force from this (imposed) loading is

$$V = \frac{wL}{2} + \frac{P}{2}$$

$$V = \frac{1.5 \times 20 \times 10}{2} + \frac{1.5 \times 15}{2} = 161.25 \, \text{kN}$$

The shear force at midspan is

$$V = \frac{P}{2} = \frac{1.5 \times 15}{2} = 11.25 \, \text{kN}$$

And the shear force diagram from imposed loads only is sketched in Figure 6.20.

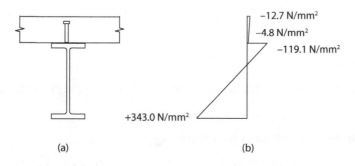

Figure 6.19 Stress distribution under ULS loads. (a) Cross section and (b) stress distribution.

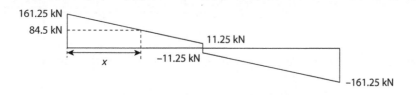

Figure 6.20 Shear force diagram (from imposed loads only).

From Equation 6.12, the shear flow at the supports is

$$\text{Shear flow} = \frac{VA'\bar{y}}{I}$$

$$\text{Shear flow} = \frac{161.25 \times 10^3 \times 50 \times 130 \times (209.5 - 65)}{475.9 \times 10^6} = 318.2 \text{ N/mm}$$

From Equation 6.13, the maximum shear stud spacing at the ends of the beam is

$$\text{Spacing} = \frac{\text{stud strength}}{\text{shear flow}}$$

$$\text{Spacing} = \frac{50 \times 10^3}{318.2} = 157.1 \text{ mm}$$

5. If the 50 kN studs are spaced at 300 mm, the corresponding shear flow (Equation 6.13) is

$$\text{Shear flow} = \frac{50 \times 10^3}{300} = 166.7 \text{ N/mm}$$

Rearranging Equation 6.12 provides the shear strength

$$V = \frac{\text{shear flow} \times I}{A'\bar{y}}$$

$$V = \frac{166.7 \times 475.9 \times 10^6}{50 \times 130 \times (209.5 - 65)} \times 10^{-3} = 84.5 \text{ kN}$$

The distance (x; see Figure 6.20) from the end of the beam, where the shear force is 84.5 kN, is

$$x = \frac{161.25 - 84.5}{1.5 \times 20} = 2.558 \text{ m}$$

Therefore, the stud spacing can be increased to 300 mm, 2.558 m from the supports.

Example 6.10: Primary beam in a building

The framing arrangement for the floor of a building is shown in Figure 6.21. A regular column grid layout of 9 by 9 m has been used throughout. All of the beams are unpropped and the slab is cast on profiled metal deck sheeting, with the direction of the span indicated by the arrows. Using the basic data below and in Figure 6.21, evaluate the following for *Beam A*:

1. Determine the bending stress under the ULS wet weight of the concrete and construction imposed loads of 0.75 kN/m².
2. Determine the midspan deflection due to dead loads and an imposed load of 5 kN/m².
3. Prove that the beam remains elastic under dead and imposed SLS loads.

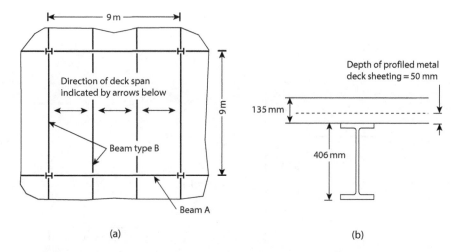

Figure 6.21 Composite beam design for a building. (a) Plan view of beam layout and (b) section through Beam A.

4. Determine if the plastic moment capacity is adequate to resist the ULS moment.
5. Determine the maximum stud spacing at the ends of the beam using plastic theory.

Basic data
Yield stress = 355 N/mm²
Young's modulus for steel = 210,000 N/mm²
Crushing strength of concrete = 40 N/mm²
Young's modulus for concrete = 20,000 N/mm²
Cross-sectional area of Beam A = 9313 mm²
Second moment of area of Beam A = 260×10⁶ mm⁴
Shear stud capacity = 85 kN per stud
Dead load of concrete slab = 2.5 kN/m²
Beam A self-weight = 0.64 kN/m
Beam B self-weight = 0.40 kN/m

1. The ULS unit loading of the slab and construction imposed loads is

$$w = 1.35g_k + 1.5q_k = 1.35 \times 2.5 + 1.5 \times 0.75 = 4.5 \text{ kN/m}^2$$

The area of slab transferred as point loads onto Beam A = 9×3 = 27 m² (see Figure 6.22).
The point loads also include the self-weight from beams type B (0.4 kN/m); therefore,

$$P = 27 \times 4.5 + 1.35 \times 0.4 \times 9 = 126.4 \text{ kN}$$

Beam A also supports a UDL developed by its own self-weight of 0.64 kN/m

$$w = 1.35 \times 0.64 = 0.86 \text{ kN/m}$$

The midspan moment is

$$M = \frac{PL}{3} + \frac{wL^2}{8}$$

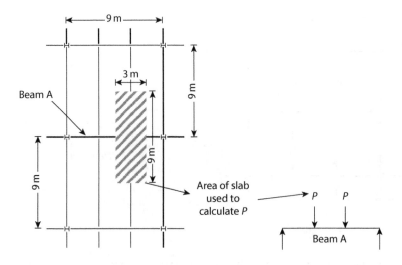

Figure 6.22 Calculation of point loads.

$$M = \frac{126.4 \times 9}{3} + \frac{0.86 \times 9^2}{8} = 387.9 \text{ kN.m}$$

The stress in the top and bottom of the beam under wet concrete is

$$\sigma = \frac{M \times (\pm)d/2}{I} = \frac{387.9 \times 10^6 \times (\pm)203}{260 \times 10^6} = \pm 302.9 \text{ N/mm}^2$$

This is less than the yield stress of 355 N/mm²; therefore, this check is passed.

2. The effective width of the slab is the lesser of either the beam spacing (3 m) or

$$b_{\text{eff}} = \frac{9000}{8} \times 2 = 2250 \text{ mm}$$

From Equation 6.2, the width of the top flange of the transformed section is

$$b_{\text{flange}} = \frac{E_c}{E_s} \times 2250 = 214.3 \text{ mm}$$

The distance from the top of the slab to the elastic neutral axis (x) is determined by taking moments about the top of the slab. Only the concrete in the top of the slab is included in the calculation, because the concrete in the troughs of the profiled metal decking is ignored to keep the analysis simple. The top of the slab is 85 mm thick (i.e. 135 mm – 50 mm; see Figure 6.21b); therefore,

$$214.3 \times 85 \times \frac{85}{2} + 9313 \times (135 + 406 \times 0.5) = x(214.3 \times 85 + 9313)$$

$$x = 142.5 \text{ mm}$$

Using the parallel axis theorem, the second moment of area of the composite beam is

$$I_{comp} = \frac{214.3 \times 85^3}{12} + 260 \times 10^6 + 9313 \times (135 + 406 \times 0.5 - 142.5)^2$$

$$+214.3 \times 85 \times (142.3 - 85 \times 0.5)^2$$

$$I_{comp} = 808.3 \times 10^6 \, mm^4$$

The unfactored dead load of the slab and the self-weight of Beam B is

$$P = 2.5 \times 9 \times 3 + 9 \times 0.40 = 71.1 \, kN$$

This point load is applied at third span points to Beam A (see Figure 6.22), in addition to the UDL from the self-weight of Beam A, 0.64 kN/m. The dead load deflection is calculated using the second moment of area of the bare steel section, since this is unpropped construction, i.e.,

$$\Delta_{dl} = \frac{5wL^4}{384EI} + \frac{23PL^3}{648EI}$$

$$\Delta_{dl} = \frac{5 \times 0.64 \times 9000^4}{384 \times 210000 \times 260 \times 10^6} + \frac{23 \times 71.1 \times 10^3 \times 9000^3}{648 \times 210000 \times 260 \times 10^6} = 34 \, mm$$

Since the imposed loads are transferred first through the secondary beams (marked B), the 5 kN/m² imposed loads develop point loads of

$$P = 9 \times 3 \times 5 = 135 \, kN \tag{6.14}$$

The midspan deflection due to the imposed loading is calculated using the second moment of area of the composite beam, i.e.,

$$\Delta = \frac{23PL^3}{648EI}$$

$$\Delta_{il} = \frac{23 \times 135 \times 10^3 \times 9000^3}{648 \times 210000 \times 808.3 \times 10^6} = 21 \, mm$$

Finally, the total deflection is

$$\Delta_{total} = 34 + 21 = 55 \, mm$$

3. The objective here is to see if the beam remains elastic under working loads. The dead load midspan bending moment is

$$M = \frac{wL^2}{8} + \frac{PL}{3}$$

$$M = \frac{0.64 \times 9^2}{8} + \frac{71.1 \times 9}{3} = 219.8 \, kN.m$$

And the corresponding stress in the outer fibres of the steel section is

$$\sigma = \frac{219.8 \times 10^6 \times (\pm 203)}{260 \times 10^6} = \pm 171.6 \ \text{N/mm}^2$$

The imposed load (Equation 6.14) develops the following midspan moment

$$M = \frac{135 \times 9}{3} = 405 \ \text{kN.m}$$

And maximum stress occurs in the bottom of the steel section, i.e.,

$$\sigma = +171.6 + \frac{405 \times 10^6 \times (135 + 406 - 142.5)}{808.3 \times 10^6} = +371.3 \ \text{N/mm}^2$$

This is more than the yield stress (355 N/mm²); therefore, the steel beam will yield under serviceability limit state loading and the section has failed this check.

4. From Equation 6.4, the tensile strength of the steel section is

$$T = f_y \times A = 355 \times 9313 \times 10^{-3} = 3306 \ \text{kN}$$

The neutral axis depth from Equation 6.8 is

$$x = \frac{3306 \times 10^3}{0.567 \times 40 \times 2250} = 64.8 \ \text{mm}$$

Since x is well above the profile metal decking, Equation 6.8 is correct for this situation. From Equation 6.9, the lever-arm depth is

$$z = \frac{d_{\text{steel}}}{2} + d_{\text{slab}} - \frac{x}{2}$$

$$z = \frac{406}{2} + 135 - \frac{64.8}{2} = 305.6 \ \text{mm}$$

From Equation 6.10, the plastic moment capacity is

$$M_{\text{pl, Rd}} = T \times z$$

$$M_{\text{pl, Rd}} = 3306 \times 305.6 \times 10^{-3} = 1010 \ \text{kN.m}$$

The ULS moment now needs determining. The point loads from beams B are

$$P = 1.35 \times (2.5 \times 9 \times 3 + 9 \times 0.40) + 1.5 \times 9 \times 3 \times 5 = 298.5 \ \text{kN}$$

And the UDL from the self-weight of Beam A is

$$w = 1.35 \times 0.64 = 0.864 \ \text{kN/m}$$

And the corresponding midspan moment is

$$M = \frac{0.864 \times 9^2}{8} + \frac{298.5 \times 9}{3} = 904.2 \text{ kN.m}$$

Since the applied moment (904.2 kN.m) is less than the moment capacity (1010 kN.m), the beam passes this check.

5. In this example, the plastic neutral axis lies in the slab ($x = 64.8$ mm); therefore, the shear studs in each third span of the beam need to develop a shear force equal to the tensile strength of the steel section, which was calculated as 3306 kN. The stud capacity is 85 kN and the minimum number of studs is

$$\text{Number of studs} = \frac{3306}{85} = 39$$

And the maximum shear stud spacing is

$$\text{Spacing} = \frac{9000/3}{39} = 77 \text{ mm}$$

Example 6.11: Bridge beam with point load

Figure 6.23a shows a cross section through a steel–concrete composite bridge spanning 40 m. The girders are unpropped whilst supporting wet concrete, although the imposed loads are resisted by composite action. Yield stress = 355 N/mm², Young's modulus for steel = 210,000 N/mm² and Young's modulus for concrete = 25,000 N/mm².

1. Determine the position of the neutral axis and the second moment of area of a single composite beam.
2. The full SLS dead load under the weight of the wet concrete is 10 kN/m (per girder). The girders were then subjected to a 12 kN/m (per girder) SLS imposed load after the concrete hardened. Determine the midspan deflection under this combined loading.
3. Determine what uniformly distributed imposed load would result in 335 N/mm² of tensile stress in the steel girders, when combined with a ULS dead load of 13.5 kN/m.
4. A combination of ULS imposed loads (i.e. with load factors applied) are shown in Figure 6.23b. Sketch the corresponding shear force diagram.
5. The shear stud capacity is 50 kN per stud. Determine the maximum shear stud spacing at the support labelled a in Figure 6.23b using elastic theory.
6. Determine the distance from the end of the beam at which the shear stud spacing can be increased to 250 mm.

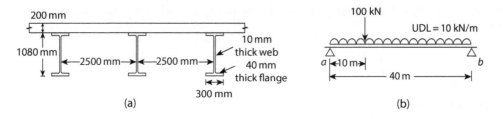

(a) (b)

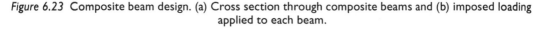

Figure 6.23 Composite beam design. (a) Cross section through composite beams and (b) imposed loading applied to each beam.

1. The second moment of area of the girder is

$$I = \frac{300 \times 1080^3 - 290 \times 1000^3}{12} = 7326 \times 10^6 \, \text{mm}^4$$

And the cross-sectional area = 34,000 mm²

The effective width of the composite beam is the lesser of either the beam spacing (2500 mm) or

$$b_{\text{eff}} = 2 \times \frac{L}{8} = \frac{2 \times 40}{8} = 10 \, \text{m}$$

From Equation 6.2, the width of top flange of the transformed section is

$$b_{\text{flange}} = \frac{2500 \times 25000}{210000} = 297.6 \, \text{mm}$$

The neutral axis is located by taking moments of area about the top of the slab

$$297.6 \times 200 \times \frac{200}{2} + 34000 \times \left(200 + \frac{1080}{2}\right) = x \times (297.6 \times 200 + 34000)$$

$$x = 332.7 \, \text{mm}$$

And by using the *parallel axis theorem,*

$$I_{\text{comp}} = 297.6 \times 200 \times \left(332.7 - \frac{200}{2}\right)^2 + \frac{297.6 \times 200^3}{12} + 34000\left(200 + \frac{1080}{2} - 332.7\right)^2$$

$$+ 7326 \times 10^6$$

$$I_{\text{comp}} = 16388 \times 10^6 \, \text{mm}^4$$

2. The deflection during the casting stage is calculated using the plain steel section second moment of area

$$\Delta_{\text{dl}} = \frac{5 \times 10 \times 40000^4}{384 \times 210000 \times 7326 \times 10^6} = 217 \, \text{mm}$$

And the deflection due to the imposed loads is calculated using the composite second moment of area

$$\Delta_{\text{il}} = \frac{5 \times 12 \times 40000^4}{384 \times 210000 \times 16388 \times 10^6} = 116 \, \text{mm}$$

And the total deflection is

$$\Delta_{\text{total}} = 217 + 116 = 333 \, \text{mm}$$

This large deflection can be partly mitigated by pre-cambering the beam to remove the 217 mm of dead load deflection.

3. The midspan moment due to the dead load is

$$M = \frac{13.5 \times 40^2}{8} = 2700 \text{ kN/m}$$

which produces the following tensile stress in the section:

$$\sigma = \frac{M \times y}{I} = \frac{2700 \times 10^6 \times (\pm)540}{7326 \times 10^6} = 199.0 \text{ N/mm}^2$$

If the imposed load per m length is w, then the corresponding moment is

$$M = \frac{w \times 40^2}{8} = 200w$$

The maximum tensile stress in the bottom flange due to imposed loading is

$$\sigma_{il} = \frac{200w \times 10^6 \times (1080 + 200 - 332.7)}{16388 \times 10^6} = 11.6w$$

If the maximum stress is 335 N/mm² this becomes

$$199.0 + 11.6w = 335$$

Therefore, the maximum UDL is 11.7 kN/m

4. The reactions at supports a and b are determined by taking moments, i.e.,

$$R_a = \frac{10 \times 40}{2} + \frac{100 \times 30}{40} = 275 \text{ kN}$$

$$R_b = \frac{10 \times 40}{2} + \frac{100 \times 10}{40} = 225 \text{ kN}$$

which are used to define the shear force diagram shown in Figure 6.24.

5. From Equation 6.12, the shear flow at support a is

$$\text{Shear flow} = \frac{VA'\bar{y}}{I}$$

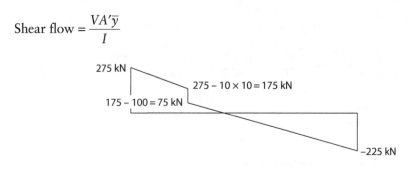

Figure 6.24 Shear force diagram.

$$\text{Shear flow} = \frac{275\times10^3 \times 297.6\times200\times(332.7-100)}{16388\times10^6} = 232.4 \text{ N/mm}$$

From Equation 6.13, the maximum shear stud spacing is

$$\text{Spacing} = \frac{50\times10^3}{232.4} = 215 \text{ mm}$$

6. The shear flow at 250 mm centre spacing is

$$\text{Shear flow} = \frac{50\times10^3}{250} = 200 \text{ N/mm}$$

Rearranging Equation 6.12 provides the shear strength

$$V = \frac{200\times16388\times10^6}{297.6\times200\times(332.7-100)} \times 10^{-3} = 236.6 \text{ kN}$$

The distance (x) from the support at which this shear force occurs is

$$x = \frac{275-236.6}{10} = 3.84 \text{ m}$$

Problems

Solutions to these problems are provided at https://www.crcpress.com/9781498741217

P.6.1. Calculate the effective width for a composite beam spanning 12 m between simple supports. The beams are spaced at 3.5 m centres.
Ans. 3.0 m.

P.6.2. A building floor comprises simply supported beams spanning 10 m and spaced at 2.5 m centres. The beams are unpropped during construction and the slab uses profiled metal decking, as shown in Figure 6.25. Young's modulus for the concrete and steel is 13,667 N/mm² and 210,000 N/mm², respectively, and the steel section has a cross-sectional area of 7600 mm² and a second moment of area of 21,508 cm⁴.

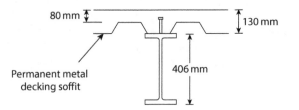

Figure 6.25 Composite beam with profiled metal decking.

a. Determine the maximum stress in the beam under the ULS wet weight of the slab (1.92 kN/m² unfactored) + the steel section self-weight (0.52 kN/m unfactored) + the load of the construction workers and equipment (0.75 kN/m² unfactored).

b. Determine the position of the neutral axis and the second moment of area of the uncracked composite slab.

c. The composite beams are designed to support an imposed load of 6 kN/m², in addition to the concrete self-weight (1.92 kN/m²) and the steel section self-weight (0.52 kN/m). Determine the distribution of stresses under these unfactored loads.

d. Determine the total deflection at the serviceability limit state under combined dead and imposed loads.

Ans. (a) ±118 N/mm², (b) x = 148 mm, I_{comp} = 634 × 10⁶ mm⁴, (c) top of the slab = −2.85 N/mm², bottom of the slab = −0.35 N/mm², top of the steel section = −68.1 N/mm², bottom of section = +177.5 N/mm² and (d) 30 mm.

P.6.3. A building floor comprises simply supported beams spanning 8 m between simple supports. The beams are spaced at 3.5 m centres. The beams support the full weight of the floor slab, which is 150 mm deep, an imposed load of 4 kN/m² and a centrally applied point (imposed) load of 20 kN per beam. The beams and the slab are joined together to act compositely and they are unpropped during construction. Determine the following:

a. The beam stresses and deflections when resisting the unfactored dead weight of the concrete only.

b. The position of the neutral axis and the second moment of area of the uncracked composite slab.

c. Sketch the stress distribution under working loads in the composite beam.

d. The total deflection under working loads.

Basic data
Second moment of area of the steel beam = 15,000 cm⁴
Cross-sectional area of steel section = 6000 mm²
Beam weight = 0.5 kN/m, concrete = 25 kN/m³
Depth of the steel section = 400 mm
Young's modulus for the steel = 210,000 N/mm²
Young's modulus for the concrete = 25,000 N/mm²

Ans. (a) Maximum stress = ±145 N/mm², deflection = 23.0 mm, (b) x = 114.6 mm, I_{comp} = 605 × 10⁶ mm⁴, (c) top of the slab = −3.4 N/mm², bottom of the slab = 1.1 N/mm², top of section = −136 N/mm², bottom of section = 254 N/mm² and (d) 30.6 mm.

P.6.4. A building floor comprises simply supported beams spanning 9 m between simple supports and spaced at 4 m centres. The total unfactored dead load is 2.05 kN/m² (inclusive of beam self-weight) and the imposed load is 6 kN/m². The beam and the slab are joined together to act compositely and the beam is unpropped during construction. The 130 mm deep slab uses profiled metal decking with a trough depth of 50 mm. Determine:

a. The bending stresses and deflection under the unfactored weight of the wet concrete only

b. The position of the neutral axis and the second moment of area of the uncracked composite beam

c. The stress distribution under working loads in the composite beam

    d. The total deflection under working loads in the composite stage
    e. The plastic moment capacity of the composite beam

Basic data

    Second moment of area of the steel beam = 21,508 cm^4
    Cross-sectional area of steel section = 7600 mm^2
    Depth of the steel section = 406 mm
    Yield stress = 275 N/mm^2
    Concrete cylinder crushing stress = 40 N/mm^2
    Young's modulus for the steel = 210,000 N/mm^2
    Young's modulus for the concrete = 25,000 N/mm^2

Ans. (a) ±78 N/mm^2, 16 mm, (b) 116.7 mm, 708 × 10^6 mm^4, (c) top of the slab = −4.8 N/mm^2, top of profiled metal decking = −1.5 N/mm^2, top of section = −73.4 N/mm^2, bottom of section = 222 N/mm^2, (d) 29.3 mm and (e) 653.1 kN.m.

P.6.5. Figure 6.26 shows a beam supporting a uniformly distributed imposed load of 6 kN/m^2 (unfactored) in addition to a centrally applied imposed load of 30 kN. The beam is to be designed elastically and will use unpropped construction, with a 10 m span and beams spaced at 2.5 m centres.

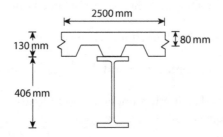

Figure 6.26 Cross section through a composite beam.

    a. Determine the stress distribution in the steel beam under ULS wet weight of the concrete.
    b. Determine the second moment of area of the composite section.
    c. Determine the elastic stress distribution in the central part of the beam under full ULS dead and imposed loads.
    d. The beam is constructed using 70 kN capacity shear connectors. Determine the maximum spacing of the shear connectors at the end of the beam using elastic design theory.

Basic data

    Second moment of area of the steel beam = 21,508 cm^4
    Cross-sectional area of steel section = 7600 mm^2
    Beam weight = 0.52 kN/m
    Slab weight = 1.92 kN/m^2
    Depth of the steel section = 406 mm
    Young's modulus for the steel = 210,000 N/mm^2
    Young's modulus for the concrete = 13,667 N/mm^2

Ans. (a) ±85 N/mm^2, (b) 634 × 10^6 mm^4, (c) top of the slab = −6.0 N/mm^2, top of profiled metal decking = −2.7 N/mm^2, top of section = −95 N/mm^2, bottom of section = 327 N/mm^2 and (d) 238 mm.

Chapter 7

Reinforced concrete beams and columns

This chapter introduces the basic theory governing the design of RC members subjected to bending and compression. The design of beams in bending and shearing is considered, as well as the basic applications of mechanics to the detailing of reinforcement and the calculation of the minimum area of reinforcement. The control of deflections is not considered, because the stiffness of reinforced concrete (RC) beams is difficult to assess due to cracking, which is normal in RC beams. Because of this, deflections are normally controlled by empirical methods involving span-to-depth ratios. This chapter then moves on to consider the design of columns and shows how to construct moment versus axial force design charts. These include charts for the design of non-rectangular and asymmetric columns, as well as columns subjected to biaxial bending.

7.1 MATERIAL PROPERTIES

RC is a composite of steel and concrete. Both materials have the same value of coefficient of linear expansion (12 microstrain per °C), although this is the only property that is the same. The most important difference is that concrete is brittle, whereas steel is ductile.

There are two ways to test concrete strength, cylinder tests and cube tests. Table 7.1 shows the compression strengths for a range of different concrete grades. A typical designation for a grade of concrete is 'C25/30'. In this case, the '25' refers to the compression strength determined from cylinder tests and the '30' refers to the cube strength. Cube tests give higher strengths than cylinders because of the friction between the test machine and the cube. This book considers calculations based on cylinder strengths, since that is the approach used by the Eurocodes.

Table 7.1 Strength grades for concrete

Concrete grade	C25/30	C30/37	C35/45	C40/50	C50/60
Cylinder test 28-day compressive strength, f_{ck} (N/mm²)	25	30	35	40	50
Short-term Young's modulus, E_c (N/mm²)	31,000	33,000	34,000	35,000	37,000
Tensile strength, f_{ctm} (N/mm²)	2.6	2.9	3.2	3.5	4.1

The design compressive strength (f_{cd}) of concrete is

$$f_{cd} = \frac{0.85 f_{ck}}{\gamma_c}$$

$$f_{cd} = 0.567 f_{ck} \tag{7.1}$$

where
  f_{ck} is the crushing strength (see Table 7.1).
  γ_c is the partial safety factor, set at 1.5.

Concrete is brittle and it is assumed to fail after an ultimate strain of only 0.0035 (0.35%). This affects many calculations, as does the design strength (f_{yd}) of rebar:

$$f_{yd} = \frac{f_{yk}}{\gamma_s}$$

$$f_{yd} = 0.87 f_{yk} \tag{7.2}$$

where
  f_{yk} is the yield stress.
  γ_s is the partial safety factor, which is set at 1.15.

It is often necessary to check to see if rebar yields before the concrete has failed in crushing. The yield strain (ε_y) is needed for these calculations. The yield strength of standard grade rebar is 500 N/mm²; therefore

$$\varepsilon_y = \frac{f_{yk}}{\gamma_s E_s} = \frac{500}{1.15 \times 210000} = 0.002 \tag{7.3}$$

Short-term values for Young's modulus (E_c) are also shown in Table 7.1. Inelastic movements, which occur over months and years, known as *creep*, reduce Young's modulus to less than half of these values when resisting permanent loads. The tensile strength (f_{ctm}) of concrete is another important parameter. It is approximately 1/10th of the crushing strength, although more accurate values are shown in Table 7.1. It is worth noting that the tensile strength of concrete is not reliable and should not be used to support loads.

The cross-sectional area of groups of standard diameter rebar are shown in Table 7.2. The number and diameter of rebar used is written in shorthand as the number, followed by the letter *H* to designate high yield and the diameter. For example, *2H32* designates two high yield rebars with a diameter of 32 mm.

Dead weight features in many calculations. In this book, the density of RC is taken as 25 kN/m³, inclusive of rebar weight.

Table 7.2 Sectional areas of groups of rebar, in mm²

Diameter (mm)	Number of bars									
	1	2	3	4	5	6	7	8	9	10
6	28	57	85	113	141	170	198	226	254	283
8	50	101	151	201	251	302	352	402	452	503
10	79	157	236	314	393	471	550	628	707	785
12	113	226	339	452	565	679	792	905	1018	1131
16	201	402	603	804	1005	1206	1407	1608	1810	2011
20	314	628	942	1257	1571	1885	2199	2513	2827	3142
25	491	982	1473	1963	2454	2945	3436	3927	4418	4909
32	804	1608	2413	3217	4021	4825	5630	6434	7238	8042
40	1257	2513	3770	5027	6283	7540	8796	10053	11310	12566

7.2 MOMENT CAPACITY OF BEAMS

The basic notation for an RC beam is shown in Figure 7.1a. The hoop-shaped rebar is a shear link and these resist shear forces. When a beam flexes, the concrete in tension will crack, and can be assumed for design purposes to carry no tensile stress. If the rebar yields, then the tensile force in the rebar is

$$T = 0.87 f_{yk} A_s \tag{7.4}$$

where
A_s is the cross-sectional area of tension steel (see Figure 7.1a).

The rebar in the compression zone will also develop compression stresses. If the rebar has yielded, then the compression force in the rebar (C_s) is

$$C_s = 0.87 f_{yk} A_s' \tag{7.5}$$

where
A_s' is the area of compression steel (Figure 7.1a).

The concrete develops a compression force, shown as C_c in Figure 7.1b. If the crushing strain in the concrete is reached ($\varepsilon_c = 0.0035$ as shown in Figure 7.1c), then a uniform compression

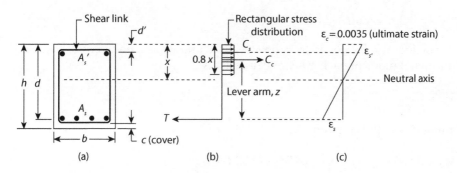

Figure 7.1 Notation, forces and strains in a RC beam at the point of failure. (a) Notation, (b) design forces and (c) strains.

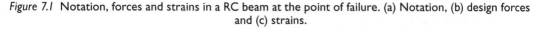

stress distribution is assumed to exist in the compression part of the beam, as sketched in Figure 7.1b. The compression zone is assumed to extend from the outer edge of the beam to a depth of $0.8x$, where x is the neutral axis depth. The total force in the concrete (C_c) is the design crushing stress (Equation 7.1), multiplied by the area of the compression block, which is the width of the compression half of the beam (b) multiplied by the depth of the stress block ($0.8x$), that is:

$$C_c = 0.567 \times f_{ck} \times b \times 0.8x \tag{7.6}$$

From force equilibrium

$$T = C_c + C_s \tag{7.7}$$

It is very important to ensure that beams are ductile. This is achieved by ensuring that failure involves yielding of the tension rebar, which is ductile, rather than crushing of the concrete, which is brittle. Ductility is achieved indirectly by limiting the depth to the neutral axis (x) to

$$x \leq 0.45 \times d \tag{7.8}$$

where
   d is known as the *effective depth* and is illustrated in Figure 7.1a, where

$$d = h - c - \phi_s - \frac{\phi_b}{2} \tag{7.9}$$

h is the beam depth.
C is known as the 'cover'; see Figure 7.1a.
ϕ_s is the diameter of the shear links.
ϕ_b is the diameter of the tension rebar.

If the beam is in equilibrium, then the applied moment (M_{Ed}) is resisted by a couple between these opposing forces. The lever arm distance, z (Figure 7.1b), is the distance between C_c and T:

$$z = d - 0.4x \tag{7.10}$$

If x is at the limit of $0.45d$, then

$$z = d - 0.4 \times (0.45d) = 0.82d \tag{7.11}$$

If the beam does not have compression steel, it is known as *singly reinforced* and the maximum (ultimate) moment such a beam can resist is

$$M_u = C_c z \tag{7.12}$$

Combining this with Equations 7.6, 7.8 and 7.11

$$M_u = 0.567 f_{ck} \times b \times 0.8 \times (0.45d) \times (0.82d)$$

$$M_u = 0.167 f_{ck} b d^2 \tag{7.13}$$

This is the maximum moment a beam can resist safely, without compression steel. The addition of compression steel lowers the depth to the neutral axis. Therefore, compression steel is used if the applied moment, $M_{Ed} > M_u$. These beams are termed *doubly reinforced* and the first task is to determine if a beam is to be singly or doubly reinforced.

7.2.1 Singly reinforced beams

The moment capacity of beams without compression steel can be determined by taking moments about the centre of the concrete compression force, C_c (see Figure 7.1b):

$$M = T \times z \tag{7.14}$$

Combining this with Equation 7.4 provides the area of tension steel needed to resist the applied moment (M_{Ed})

$$A_s = \frac{M_{Ed}}{0.87 f_{yk} z} \tag{7.15}$$

The lever-arm distance, z, needs to be calculated. This involves the solution of a quadratic equation using the following formulae:

$$z = d\left(0.5 + \sqrt{0.25 - 3k/3.4}\right) \tag{7.16}$$

where

$$k = \frac{M_{Ed}}{f_{ck} b d^2} \tag{7.17}$$

Example 7.1: Singly reinforced beam

Determine the area of rebar needed to resist the midspan bending moments, for a 275 mm wide by 450 mm deep beam, which supports a 22 kN/m imposed load. The beam is simply supported and spans 7 m, f_{ck} is 35 N/mm² and 25 mm cover is provided.

The first step is to calculate the effective depth. In this calculation, it is assumed conservatively that 8 mm diameter shear links and a 40 mm tension rebar are used. From Equation 7.9

$$d = h - c - \phi_s - \frac{\phi_b}{2}$$

$$d = 450 - 25 - 8 - 40/2 = 397 \text{ mm}$$

The maximum moment a beam can resist without compression steel (Equation 7.13) is

$$M_u = 0.167 f_{ck} b d^2$$

$$M_u = 0.167 \times 35 \times 275 \times 397^2 \times 10^{-6} = 253 \text{ kN.m}$$

From Equation 1.3, the ULS load is

$$w_{\mathrm{uls}} = 1.35g_k + 1.5q_k$$

$$w = 1.35 \times 0.275 \times 0.45 \times 25.0 + 1.5 \times 22 = 37 \text{ kN/m}$$

where 25.0 kN/m³ is the density of the concrete. The corresponding midspan moment is

$$M_{\mathrm{Ed}} = \frac{37 \times 7^2}{8} = 227 \text{ kN.m}$$

Since the applied moment $M_{\mathrm{Ed}} \leq M_u$, this beam can be designed as singly reinforced and from Equation 7.17

$$k = \frac{M_{\mathrm{Ed}}}{f_{ck}bd^2}$$

$$k = \frac{227 \times 10^6}{35 \times 275 \times 397^2} = 0.150$$

And the lever arm from Equation 7.16 is

$$z = d\left(0.5 + \sqrt{0.25 - 3k/3.4}\right)$$

$$z = 397(0.5 + \sqrt{0.25 - 3 \times 0.150/3.4}) = 335 \text{ mm}$$

And the area of tension steel from Equation 7.15 is

$$A_s = \frac{M_{\mathrm{Ed}}}{0.87 f_{yk} z}$$

$$A_s = \frac{227 \times 10^6}{0.87 \times 500 \times 335} = 1558 \text{ mm}^2$$

Inspection of Table 7.2 shows two 32 mm diameter rebars (in shorthand, this is 2H32) provide 1608 mm².

Example 7.2: Investigate the effect neutral axis depth has on ductility

Two beams are identical apart from area of reinforcement used (see Figure 7.2). The beams are tested to failure, with the load deflection response shown in Figure 7.3. Use theory to calculate the depth to the neutral axis for each beam and comment on its effect on ductility, as shown in the load deflection graphs in Figure 7.3. The rebar yield stress is 500 N/mm² and the concrete crushing stress is 40 N/mm².

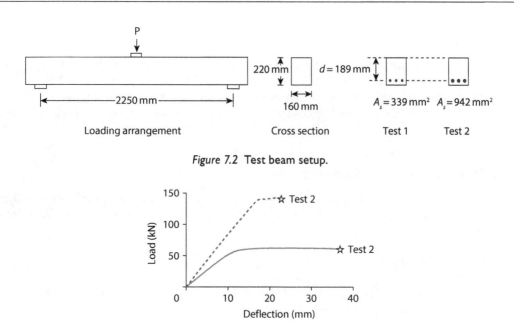

Figure 7.2 Test beam setup.

Figure 7.3 Experimental load versus deflection for beams in Figure 7.2.

Test 1: From Equation 7.4, the design tensile strength of the rebar is

$$T = 0.87 f_{yk} A_s$$

$$T = 0.87 \times 339 \times 500 \times 10^{-3} = 147.4 \text{ kN} \tag{7.18}$$

From Equation 7.6

$$C_c = 0.567 \times f_{ck} \times b \times 0.8x$$

$$C_c = 0.567 \times 40 \times 160 \times (0.8x) \times 10^{-3} = 2.9x \tag{7.19}$$

From horizontal equilibrium

$$C_c = T \tag{7.20}$$

Combining Equations 7.18, 7.19 and 7.20, the neutral axis depth is

$$x = \frac{147.4}{2.9} = 51 \text{ mm}$$

Test 2: From Equation 7.4

$$T = 0.87 \times 942 \times 500 \times 10^{-3} = 409.8 \text{ kN}$$

and it follows that

$$x = \frac{409.8}{2.9} = 141 \text{ mm}$$

As discussed earlier, $x \leq 0.45d$ in order to ensure sufficient ductility and in this case $0.45d = 85$ mm.

Test 1 is well within this limit ($x = 51$ mm) and Figure 7.3 shows that this beam deflected by almost 40 mm before reaching the collapse point. The theory suggests that Test 2 should be less ductile, since $x = 140$ mm and is therefore well over the safety limit. Figure 7.3 shows that this is indeed the case, with the collapse point occurring quickly after the elastic limit is reached.

7.2.2 Doubly reinforced beams

If the applied moment, M_{Ed}, is greater than M_u, then strength can be increased by balancing tension force with compression steel. Taking moments about the tension steel, as illustrated in Figure 7.4b

$$M_{Ed} = C_c z + C_s (d - d') \tag{7.21}$$

The neutral axis depth, x, is set at the maximum limit of $0.45d$; therefore, from Equation 7.11 and 7.6 the compression force in the concrete becomes

$$C_c = 0.567 f_{ck} \times b \times 0.8 \times 0.45d$$

$$C_c = 0.204 f_{ck} bd \tag{7.22}$$

Substituting Equations 7.4 and 7.22 into 7.7

$$0.87 A_s f_{yk} = 0.204 f_{ck} bd + 0.87 A_s' f_{yk} \tag{7.23}$$

Substituting in Equations 7.11 and 7.13

$$A_s = \frac{M_u}{0.87 f_{yk} z} + A_s' \tag{7.24}$$

Taking moments about the centre of the tension steel (Figure 7.4b)

$$M_{Ed} = 0.204 f_{ck} bd \times 0.82d + 0.87 f_{yk} A_s' (d - d')$$

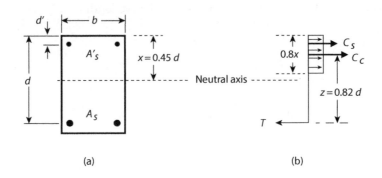

(a)　　　　　　　　　　　　　(b)

Figure 7.4 Doubly reinforced beam. (a) Cross section and (b) design forces.

Combined with Equation 7.24, this rearranges to

$$A_s' = \frac{M_{Ed} - M_u}{0.87 f_{yk}(d - d')} \tag{7.25}$$

where
 A_s' is the area.
 d' is the effective depth for the compression steel (see Figure 7.4a).

Yielding check. These calculations assumed that the reinforcement has yielded. This is a safe assumption with regard to the tension steel; however, the compression steel is located closer to the neutral axis and it may not develop sufficient strain to yield. The strain distribution at failure is shown in Figure 7.1c, and the principle of similar triangles can be applied to this to show that

$$\frac{\varepsilon_s'}{x - d'} = \frac{0.0035}{x}$$

Rearranging

$$\frac{d'}{x} = 1 - \frac{\varepsilon_s'}{0.0035}$$

From Equation 7.3, we see that normal quality steel rebar yields at a strain of 0.002; therefore

$$\frac{d'}{x} \le 1 - \frac{0.002}{0.0035} \le 0.43$$

Since the neutral axis depth was set on the limit of x = 0.45d, this becomes

$$\frac{d'}{d} \le 0.19 \tag{7.26}$$

If this holds true, then the compression steel will yield and moment calculations are correct. If not, then the compression steel will not yield before failure and Equation 7.23 will require a reduced stress for the compression steel.

Example 7.3: Doubly reinforced beam

Determine how much rebar is needed to resist the midspan moments for a 275 mm wide by 450 mm deep RC beam supporting an imposed load of 40 kN/m. The beam spans 7 m, f_{ck} is 35 N/mm² and 25 mm cover is provided.

The first step is to calculate the effective depth. Assuming 40 mm diameter main rebar and 8 mm diameter shear links are used, from Equation 7.9

$$d = 450 - 25 - 8 - 40/2 = 397 \text{ mm}$$

And the depth to the compression steel (see Figure 7.4a) is

$$d' = 25 + 8 + 20 = 53 \text{ mm}$$

From Equation 1.3 the ULS load (w) and midspan moment (M_{Ed}) are

$$w = 1.35 \times 0.275 \times 0.45 \times 25 + 1.5 \times 40 = 64 \text{ kN/m}$$

$$M_{Ed} = \frac{64 \times 7^2}{8} = 392 \text{ kN.m}$$

The ultimate moment without compression steel (Equation 7.13) is

$$M_u = 0.167 f_{ck} b d^2$$

$$M_u = 0.167 \times 35 \times 275 \times 397^2 \times 10^{-6} = 253 \text{ kN.m}$$

Since $M > M_u$, this beam requires compression steel, and from Equation 7.25

$$A_s' = \frac{M_{Ed} - M_u}{0.87 f_{yk}(d - d')}$$

$$A_s' = \frac{(392 - 253) \times 10^6}{0.87 \times 500(397 - 53)} = 929 \text{ mm}^2$$

From Equation 7.11

$$z = 0.82d = 326 \text{ mm}$$

From Equation 7.24

$$A_s = \frac{M_u}{0.87 f_{yk} z} + A_s'$$

$$A_s = \frac{253 \times 10^6}{0.87 \times 500 \times 326} + 929 = 2713 \text{ mm}^2$$

For completeness, it is necessary to check that that the compression steel yields using Equation 7.26

$$\frac{d'}{d} = \frac{53}{397} = 0.13 \leq 0.19 \therefore \text{OK}$$

Using Table 7.2, the following rebar provision can be shown to meet these requirements:

Top $\rightarrow$ 2 × 25 mm diameter rebar (shorthand 2H25)

Bottom $\rightarrow$ 2 × 40 mm + 1 No. 25 mm rebar (shorthand 2H40 + 1H25)

Example 7.4: Calculate the strength of a beam

Determine the bending strength of a 400 mm wide, 350 mm deep beam reinforced with 8 mm diameter shear links, four 32 mm diameter tension rebars, and four 20 mm diameter compression rebars. The cover provided to the shear reinforcement is 25 mm, f_{ck} is 40 N/mm² and f_{yk} is 500 N/mm².

From Equation 7.9, the effective depth to the tension steel is

$$d = 350 - 25 - 8 - 32/2 = 301 \text{ mm}$$

and the effective depth to the compression steel is

$$d' = 25 + 8 + 20/2 = 43 \text{ mm}$$

From Table 7.2, the areas of rebar are as follows: A_s is 3217 mm² and A'_s is 1257 mm². From Equation 7.4, the tensile strength of the tension rebar is

$$T = 0.87 f_{yk} A_s$$

$$T = 0.87 \times 3217 \times 500 \times 10^{-3} = 1399 \text{ kN} \tag{7.27}$$

From Equation 7.5

$$C_s = 0.87 f_{yk} A'_s$$

$$C_s = 0.87 \times 1257 \times 500 \times 10^{-3} = 547 \text{ kN} \tag{7.28}$$

From Equation 7.6

$$C_c = 0.567 f_{ck} \times b \times 0.8x$$

$$C_c = 0.567 \times 40 \times 400 \times 0.8x$$

$$C_c = 7258x \tag{7.29}$$

From Equation 7.7

$$T = C_c + C_s$$

$$1399 \times 10^3 = 7258x + 547 \times 10^3$$

which solves to x = 117 mm, and from Equation 7.29

$$C_c = 7258 \times 117 \times 10^{-3} = 849 \text{ kN}$$

Taking moments about the tension steel (see Figure 7.1b), the moment capacity is

$$M = C_s(d - d') + C_c(d - 0.4x) \tag{7.30}$$

or

$$M = 547 \times (301 - 43) \times 10^{-3} + 849 \times (301 - 0.4 \times 117.5) \times 10^{-3} = 356 \text{ kN.m}$$

7.3 THE MAXIMUM AND MINIMUM AREAS OF REINFORCEMENT IN A BEAM

It is not possible to keep increasing strength by unlimited additions of steel. For beams, a maximum of 4% of the cross section can be taken as steel.

The minimum area of steel is defined by the objective to ensure that the beam fails in a safe and controlled manner. The elastic section modulus (W_{el}) of a rectangular beam of width (b) and depth (h) is

$$W_{el} = \frac{b \times h^2}{6} \tag{7.31}$$

The moment capacity of an unreinforced concrete section is reached when the tensile strength of the concrete (f_{ctm}) is reached. Thus

$$M_{unreinforced} = f_{ctm} \times W_{el} \tag{7.32}$$

For safe design, it is necessary to ensure that the bending strength of the cracked (reinforced) beam is not less than that of the uncracked (plain) concrete beam, i.e.,

$$M_{reinforced} \geq M_{unreinforced} \tag{7.33}$$

This prevents cracking initiating a complete tensile failure of the reinforcement. If this occurs, then the beam can fail almost instantly. The moment capacity of the reinforced section is

$$M_{reinforced} = A_s f_{yk} z$$

In lightly reinforced sections, the lever arm is

$$z \approx 0.95d$$

Therefore

$$M_{reinforced} = 0.95 A_s f_{yk} d \tag{7.34}$$

Inputting Equations 7.32 and 7.34 into Equation 7.33

$$A_s f_{yk} 0.95d \geq \frac{f_{ctm} b h^2}{6}$$

or

$$A_s \geq \frac{0.18 f_{ctm} b h^2}{f_{yk} d}$$

Assuming that $h \approx 1.2d$, this becomes

$$A_s \geq \frac{0.26 f_{ctm} b d}{f_{yk}} \qquad (7.35)$$

If this minimum area of steel is insufficient to resist the loads after first cracking, then a beam may fail suddenly if overloaded. Therefore, providing more than the bare minimum is prudent in many situations. Equation 7.35 tells us that the minimum area increases as the concrete strength increases; therefore, overstrength concrete could lead to brittle failure.

Example 7.5: Calculation of the minimum area of reinforcement

Determine the minimum amount of tension reinforcement required for a 1200 mm wide beam that has an effective depth of 960 mm. The tensile (cracking) strength of the concrete is 3.5 N/mm² and $f_{yk} = 500$ N/mm².

From Equation 7.35

$$A_s \geq \frac{0.26 \times 3.5 \times 1200 \times 960}{500} = 2097 \text{ mm}^2$$

7.4 ANCHORAGE OF REINFORCEMENT AND LAPPING OF BARS

The surface of rebar is ribbed to prevent slippage through the concrete (see Figure 7.5). Despite this, rebar must be properly anchored in order to generate their design tensile or compressive forces. The anchorage length, L_b, is the minimum length required to prevent the rebar pulling out or pushing through the concrete (see Figure 7.6a). Bent bars are harder to pull out and therefore require less anchorage (see Figure 7.6b). The anchorage length is defined using codes of practice, although in simple terms, the basic anchorage length is equal to the anchorage length factor multiplied by the rebar diameter (see Table 7.3). For example, 25 mm diameter rebar in C30/37 concrete would require an anchorage length of 25 × 36 = 900 mm.

Lapping. The normal way to joint rebar is to lay one next to another and allow the concrete to transfer the force. This is known as 'lapping' and is illustrated in Figure 7.6c. Lapping can be unreliable. For example, the cover at laps can sometimes fail by splitting. Whilst almost all rebars are joined by lapping, this technique needs to be used carefully; therefore:

- Laps should spread out rather than concentrated.
- Avoid lapping more than 50% of the rebar at any one point.
- Position laps at positions of low moments.

Curtailment of tension reinforcement. In order to economise on rebar, it is usual to match the moment capacity to the bending moment diagram. This process is illustrated by Figure 7.7a, which shows the positions of the 50% midspan moment points for a beam

Figure 7.5 Rebar showing the ribbed surface designed to improve the bond with concrete.

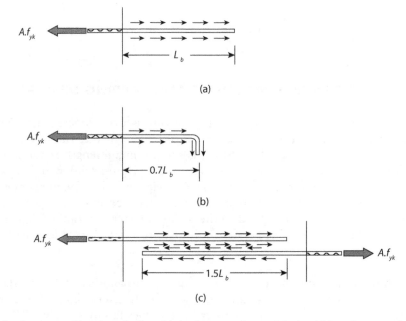

(a)

(b)

(c)

Figure 7.6 Anchorage and lapping of rebar. (a) Anchorage of a straight bar, (b) anchorage of a bent rebar and (c) lapping of two rebar.

Table 7.3 Anchorage length factors for straight rebar in tension or compression

Concrete grade	C25/30	C30/37	C35/45	C40/50
Anchorage length factor	41	36	33	30

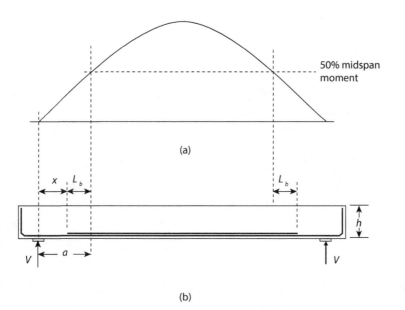

(a)

(b)

Figure 7.7 Illustration of the curtailment of tension reinforcement. (a) Bending moment diagram and (b) curtailment points for 50% of the midspan reinforcement.

supporting a UDL. Figure 7.7b shows the corresponding points where 50% of the midspan rebar can be ended. It is *very important* to anchor the rebar that continues to the supports properly. Failure to do so seriously affects the shear capacity (not just the moment capacity). Figure 7.7b shows an example of how rebar can be anchored using bars bent 90°.

Example 7.6: Curtailment of rebar

A beam spans 11 m between simple supports and is 300 mm wide and 540 mm deep. It supports its self-weight in addition to an imposed UDL of 15 kN/m and a centrally applied unfactored point load of 40 kN (dead) and 40 kN (imposed). The concrete density is 25 kN/m³ and the grade is C30/37.

1. Determine the midspan ULS moment and support shear force.
2. Determine the distance from the supports that the moment is equal to 50% of the midspan moment.
3. If four 25 mm rebars were used as midspan tension reinforcement, determine the distance from the supports such that two of the rebars (i.e. 50%) can be curtailed.

1. From Equation 1.3 the ultimate limit state UDL is

$$w = 1.35 \times 25 \times 0.3 \times 0.54 + 1.5 \times 15 = 27.9 \text{ kN/m}$$

and the point load is

$$P = 1.35 \times 40 + 1.5 \times 40 = 114 \text{ kN}$$

The midspan moment is

$$M = \frac{PL}{4} + \frac{wL^2}{8}$$

$$M = \frac{114 \times 11}{4} + \frac{27.9 \times 11^2}{8} = 736 \text{ kN.m}$$

and the support shear force is

$$V = \frac{P}{2} + \frac{wL}{2}$$

$$V = \frac{114}{2} + \frac{27.9 \times 11}{2} = 210 \text{ kN}$$

2. The 50% moment point is located at distance a from the support (see Figure 7.7b), and taking moments about this position

$$0.5M = V \times a - w \times \frac{a^2}{2}$$

$$0.5 \times 736 = 210 \times a - 27.9 \times \frac{a^2}{2}$$

The roots of this quadratic are 2.02 and 13.02 m. In other words, the 50% midspan moment is located 2.02 m from each support.

3. From Table 7.3, the anchorage length factor for C30/37 concrete is 36; therefore, the anchorage length

$$L_b = 36 \times 25 = 900 \text{ mm}$$

The distance, x, from the support to the point of curtailment (see Figure 7.7b) is

$$x = a - L_b = 2020 - 900 = 1120 \text{ mm}$$

7.5 SHEAR CAPACITY OF BEAMS

Figure 7.8a shows diagonal shear cracks propagating away from the supports at an angle of approximately 45°. Shear strength is provided by vertical rebar (known as *shear links*) that pass through the shear cracks. Similar rebars are shown being fixed in Figure 7.9. During design, the objective is to ensure that the shear resistance (V_{Rd}) is greater than the applied shear force (V_{Ed}). Concrete has an inherent shear capacity in the absence of shear links. Lightly loaded elements, such as slabs, rely on the shear strength of the concrete without the need for shear links. Shear failures tend to be sudden and dangerous; therefore, members of significance, such as *all* beams, *must* be strengthened with shear links.

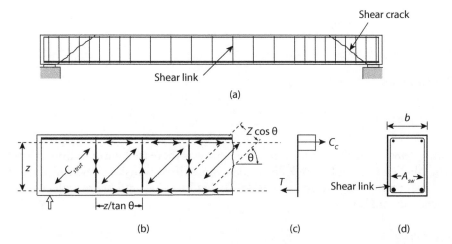

Figure 7.8 The mechanics of shear reinforcement. (a) Beam showing the rebar arrangement and shear cracks, (b) truss analogy for shear, (c) stress block for bending and (d) cross section.

Figure 7.9 A steel fixer attaching a shear link to a beam.

During the design process, the member is assumed to resist shear by truss action, illustrated in Figure 7.8b. The main tensile reinforcement comprises the bottom chord of the truss. The top chord comprises the concrete compression block developed to resist bending moments and the distance between these notional members is the lever arm distance, z. The diagonal struts are located at an angle θ and the vertical struts are spaced at a distance $z/\tan(\theta)$; see Figure 7.8b.

The vertical members in the notional truss are composed of the shear links, which have an area, A_{sw}. From force equilibrium, the shear capacity (V_{Rd}) is equal to the tensile strength of the shear links passing through a shear crack. As the shear link spacing (s) is reduced, the shear strength increases. This increase is proportional to the ratio between s and the distance between vertical members shown in Figure 7.8b. Thus, the design shear strength is

$$V_{Rd} = 0.87 f_{yk} A_{sw} \frac{z}{s \tan(\theta)} \tag{7.36}$$

where
  s is the shear link spacing.
  θ is the angle of the diagonal strut (Figure 7.8b).
  z is the lever arm depth.

If $z \approx 0.9d$ and $\theta = 45°$, then

$$V_{Rd} = 0.78 f_{yk} d \frac{A_{sw}}{s} \tag{7.37}$$

Maximum shear strength. If shear reinforcement was increased without limit, then the diagonal strut labelled C_{strut} in Figure 7.8b would eventually be crushed, and this limits the shear capacity. Diagonal cracks form, as shown in Figure 7.8a, and these reduce the crushing stress in the diagonal strut to

$$\sigma_{Rd} = \nu f_{cd}$$

where the empirical reduction factor is

$$\nu = 0.6(1 - f_{ck}/250)$$

The diagonal strut width is $z \cos(\theta)$, as illustrated in Figure 7.8b; therefore, the crushing strength is

$$C_{strut} = \sigma_{Rd} \times b \times z \cos(\theta)$$

where
  b is the width of the beam.

From force equilibrium

$$V_{Rd,\,max} = C_{strut} \sin(\theta)$$

Combining the above equations

$$V_{Rd,\,max} = 0.567 f_{ck} \times 0.6(1 - f_{ck}/250) \times b \times z \cos(\theta) \times \sin(\theta)$$

If $z \approx 0.9d$ and $\theta = 45°$, then the maximum shear strength is

$$V_{Rd,\,max} = 0.153 f_{ck} bd \left(1 - \frac{f_{ck}}{250} \right) \tag{7.38}$$

Maximum spacing of shear links. Since shear cracks tend to run at 45° to the plane of the member, it is important that shear links are not spaced so far apart that a crack can form in between the links. Therefore, the maximum spacing of the shear links is 0.75d.

Economy of shear reinforcement. The provision of shear links is a comparatively expensive operation; therefore, it is usual to match the shear strength to the shear force diagram. The method for establishing the position where the shear link spacing can be increased is explained in Example 7.7.

Example 7.7: Design of shear reinforcement

A 250 mm wide and 400 mm deep RC beam spans 7 m between simple supports. It supports its self-weight in addition to an unfactored imposed UDL of 25 kN/m and a centrally applied unfactored point load of 20 kN dead and 20 kN imposed. The strengths are f_{yk} = 500 N/mm² for the reinforcement and f_{ck} = 35 N/mm² for the concrete. The cover is 25 mm and 40 mm diameter main tension rebar are used.

1. Determine the ULS shear force.
2. Determine the maximum shear strength the beam can possess, irrespective of how much shear reinforcement is used.
3. It has been decided to use 8 mm diameter shear links in the beam. Determine the shear link spacing at the ends of the beam.
4. Determine the distance from the end supports at which the shear link spacing can be increased to 200 mm.

1. From Equation 1.3 the Ultimate limit state UDL

$$w = 1.35 \times 0.25 \times 0.4 \times 25 + 1.5 \times 25 = 40.9 \text{ kN/m}$$

The point load is

$$P = 1.35 \times 20 + 1.5 \times 20 = 57 \text{ kN}$$

and the shear force is

$$V_{Ed} = \frac{57}{2} + \frac{40.9 \times 7}{2} = 171.6 \text{ kN}$$

2. The effective depth from Equation 7.9 is

$$d = 400 - 25 - 8 - 40/2 = 347 \text{ mm}$$

From Equation 7.38, the maximum shear strength the beam can attain is

$$V_{Rd,\,max} = 0.153 f_{ck} bd \left(1 - \frac{f_{ck}}{250} \right)$$

$$V_{Rd,\,max} = 0.153 \times 35 \times 250 \times 347 \times (1 - 35/250) \times 10^{-3} = 400 \text{ kN}$$

Since this is much higher than the applied shear force, there is no problem from concrete crushing resulting in early shear failure.

3. The 8 mm diameter shear links are hoop shaped; therefore, the rebar that comprises the shear links must pass through any shear crack twice, as sketched in Figure 7.8. Because of this, A_{sw} is *twice* the cross-sectional area of a single 8 mm rebar, or 101 mm². The maximum shear stud spacing is when $V_{Rd} = V_{Ed}$ and from Equation 7.37 the shear link spacing is

$$s = 0.78 f_{yk} d \frac{A_{sw}}{V_{Rd}}$$

Figure 7.10 A RC column prior to casting.

$$s = \frac{0.78 \times 500 \times 347 \times 101}{171.6 \times 10^3} = 80 \text{ mm}$$

Therefore, provide 8 mm shear links at 80 mm centres.

4. From Equation 7.37, shear links at 200 mm spacing develop the following shear capacity

$$V_{Rd} = 0.78 \times 500 \times 347 \times \frac{101}{160} \times 10^{-3} = 68.3 \text{ kN}$$

The UDL is 40.9 kN/m, which means that for every metre away from a support the shear force drops by 40.9 kN, until the point load is reached. The shear link spacing can be increased when the shear force drops to 68.3 kN. This occurs at a distance x from the support, where

$$x = \frac{171.6 - 68.3}{40.9} = 2.53 \text{ m}$$

The shear link spacing can increase to 200 mm at a distance of 2530 mm from the supports.

7.6 INTRODUCTION TO COLUMN DESIGN

Columns can fail either by crushing or buckling or a combination of the two. The failure mode will depend on the slenderness, as illustrated by Figure 7.11, which is measured by the ratio between the effective length (L_{cr}) and the column depth (h).

The crushing strength is simply the sum of the strengths of the concrete and steel in a column, i.e.,

$$N_{crush} = f_{cd} \times A_c + f_{yd} \times A_s \tag{7.39}$$

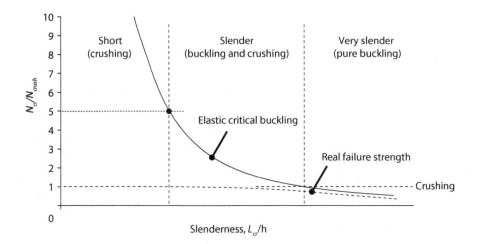

Figure 7.11 Relationship between failure mode and column slenderness.

where A_c and A_s are the cross-sectional areas of the steel and concrete, respectively. Equation 7.39 combined with Equations 7.1 and 7.2 becomes

$$N_{\text{crush}} = 0.567 f_{\text{ck}} A_c + 0.87 f_{\text{yk}} A_s \tag{7.40}$$

The elastic critical buckling force can be calculated using Euler's well-known formula, i.e.,

$$N_{\text{cr}} = \frac{\pi^2 EI}{L_{\text{cr}}^2} \tag{7.41}$$

Figure 7.11 shows the ratio $N_{\text{cr}}/N_{\text{crush}}$ plotted against slenderness. This ratio can be used to estimate whether failure will be by crushing (short columns) or by buckling (slender columns). In very simple terms, columns can be defined as short if

$$\frac{N_{\text{cr}}}{N_{\text{crush}}} > 5 \tag{7.42}$$

If this is the case, then effects of buckling are minor and can be ignored, in which case the column must be capable of resisting the combined effects of the compression and moments. The columns in most buildings are short, and therefore practical column design is simplified. Codes of practice provide more exact methods for determining if a column is short or slender, but this limiting ratio is a good first principles-based guide. The design of slender columns is outside the scope of this book.

Example 7.8: Classify a column as short or slender

A 3250 mm long column has a 400 mm square cross section and is reinforced with four rebars with a combined area of 1963 mm²; it can conservatively be assumed to be simply supported at the ends. Estimate the mode of failure, i.e., buckling or crushing, if $f_{\text{ck}} = 40$ N/mm², $f_{\text{yk}} = 500$ N/mm² and Young's modulus for the concrete, $E_c = 15{,}000$ N/mm².

The crushing strength from Equation 7.40 is

$$N_{crush} = 0.567 f_{ck} A_c + 0.87 f_{yk} A_s$$

$$N_{cr} = 0.567 \times 40 \times 400^2 \times 10^{-3} + 0.87 \times 500 \times 1963 \times 10^{-3} = 4483 \text{ kN}$$

The second moment of area of the cross section is

$$I = \frac{400^4}{12} = 2.13 \times 10^9 \text{ mm}^4$$

The elastic critical compression force from Equation 7.41 is

$$N_{cr} = \frac{\pi^2 \times 15000 \times 2.13 \times 10^9}{3250^2} \times 10^{-3} = 29854 \text{ kN}$$

And from Equation 7.42

$$\frac{N_{cr}}{N_{crush}} = \frac{29854}{4483} = 6.7$$

Since this ratio is greater than 5.0, this column is classified as short and in the absence of moments, the crushing strength will be approximately 4483 kN. Note that in the interests of simplicity, the area of concrete displaced by the rebar was ignored when calculating N_{crush}. Furthermore, the second moment of area calculation ignored the extra stiffness (EI) provided by the steel. A more detailed calculation including both effects would have yielded a ratio of 8.8 instead of 6.7.

7.7 SHORT COLUMNS SUBJECTED TO COMBINED COMPRESSION AND BENDING

If it is established that buckling will not influence strength, then the ability of the cross section to resist applied compression and moments is established without further consideration of buckling. The simplest method of assessing the ability of a column to resist a combination of axial force (N) and moment (M) is to construct a design chart, known as an *M–N interaction diagram*. If the combination of M and N lies within the design envelop, as shown in Figure 7.12a, then the column is deemed satisfactory.

7.8 M–N INTERACTION DIAGRAMS

Figure 7.12a shows the three key points (1, 2 and 3) needed to define M–N interaction diagrams. For each point, the coordinates are calculated by resolving the axial forces to determine N and by taking moments about the column centreline to determine M. If the column is symmetrical as shown in Figure 7.12a, then the diagram is also symmetrical,

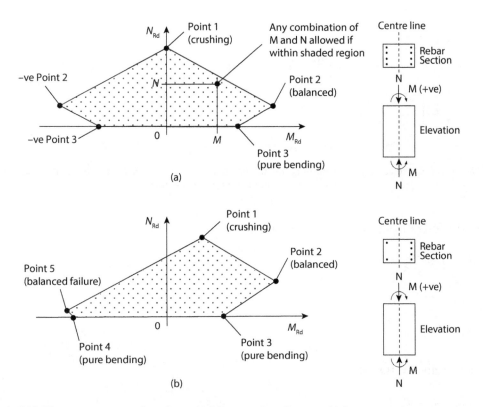

Figure 7.12 The construction of a column *M–N* interaction diagram. (a) Symmetrical cross section and (b) unsymmetrical cross section.

whereas unsymmetrical columns have unsymmetrical interaction diagrams (Figure 7.12b). If unsymmetrical, then usually the three points (1, 2 and 3) are enough for design purposes, although the process for calculating points 4 and 5 are the same as 2 and 3, except that the moment is negative, which means the tension steel becomes compression steel and *vice versa*.

The sign convention for the construction of the interaction diagrams is as follows:

1. Compression is +ve and tension is −ve.
2. The lever arm for calculation of moments is +ve if above the column centre line and −ve if below.

Point 1: Crushing. Figure 7.13 shows a column subjected to a combination of moment and compression that produces uniform crushing throughout the cross section, labelled Point 1 on Figure 7.12b. A moment is needed to balance the asymmetric forces in the rebar, since the cross section in this case is asymmetric (since $A_{s1} > A_{s2}$). From force equilibrium, the compression strength is

$$N_{Rd} = C_c + N_{s1} + N_{s2} \tag{7.43}$$

And from moment equilibrium

$$M_{Rd} = N_{s1}z_{s1} + N_{s2}z_{s2} \tag{7.44}$$

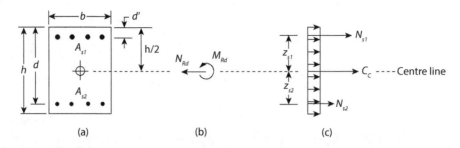

Figure 7.13 Calculation of maximum compression strength and associated moment. (a) Notation, (b) applied M and N and (c) internal forces.

where the forces from the concrete and rebar are

$$C_c = 0.567 f_{ck} \times b \times h \tag{7.45}$$

$$N_{s1} = 0.87 A_{s1} f_{yk} \tag{7.46}$$

$$N_{s2} = 0.87 A_{s2} f_{yk} \tag{7.47}$$

The lever arms for the rebar are

$$z_{s1} = h/2 - d' \tag{7.48}$$

$$z_{s2} = -(d - h/2) \tag{7.49}$$

Note that z_{s2} is negative, because the lower rebar is below the column centre line and thus produces a negative moment.

Point 2: Balanced failure. This occurs when the outermost steel begins to yield at the same time as the concrete begins to crush. Figure 7.14 shows a cross section subjected to a combination of moment and compression that produces this 'balance failure', which is labelled as Point 2 on Figure 7.12.

Force equilibrium (Equation 7.43) provides the compression strength and from moment equilibrium

$$M_{Rd} = C_c z_c + N_{s1} z_{s1} + N_{s2} z_{s2} \tag{7.50}$$

The lever arms for the rebar are provided by Equations 7.48 and 7.49. In the same manner as that used for beams, the concrete stress block is assumed to be $0.8x$ deep, where x is the depth to the neutral axis, as sketched in Figure 7.1b. The lever arm for the concrete force is

$$z_c = \frac{h}{2} - \frac{0.8x}{2} = h/2 - 0.4x \tag{7.51}$$

At balance, the strain distribution is defined by the tension steel just yielding and the concrete just beginning to crush. The yield strain for the rebar is 0.002 (see Equation 7.2), and the crushing strain for 'normal' grades of concrete is 0.0035. Using these numbers, the strain

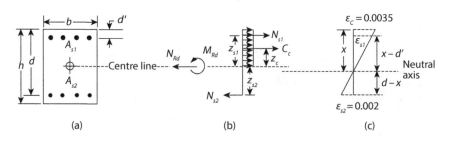

Figure 7.14 Cross section at 'balance'. (a) Notation, (b) forces and (c) strain distribution.

distribution shown in Figure 7.14c is constructed. The neutral axis depth (x) is determined using the method of similar triangles applied to this distribution, i.e.,

$$\frac{x}{0.0035} = \frac{d-x}{0.002}$$

which rearranges to

$$x = 0.636d \qquad (7.52)$$

The force in the lower rebar (N_{s2}) is determined using Equation 7.47. If the upper steel is located close to the neutral axis, it may not yield before the concrete crushes. From triangles applied to the strain distribution

$$\frac{\varepsilon_{s1}}{x-d'} = \frac{0.0035}{x}$$

which rearranges to

$$\varepsilon_{s1} = \frac{0.0035(x-d')}{x} \qquad (7.53)$$

From Equation 7.2 it was shown that if $\varepsilon_{s1} \geq 0.002$, then steel will yield, in which case the force is determined using Equation 7.46. If not, then the N_{s1} must be calculated using ε_{s1}.

Point 3: Pure bending. Since $N = 0$, the column is treated like a beam when calculating bending strength. If $A_{s1} \geq A_{s2}$ (where A_{s1} is the area of compression steel), then the concrete is not needed to balance N_{s2} (see Figure 7.15). Therefore, from moment equilibrium

$$M_{Rd} = -N_{s2}(d-d') \qquad (7.54)$$

This equation is conservative, because it ignores compression in the concrete, although this conservatism is not normally significant. If $A_{s1} < A_{s2}$, then compression in the concrete is needed to balance N_{s2} and the design method is the same as that for Point 2, except that the neutral axis depth is defined by force equilibrium from Equation 7.43, with $N_{Rd} = 0$.

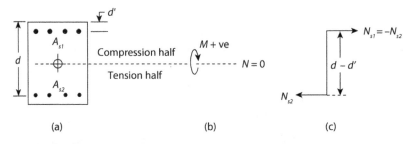

Figure 7.15 Calculation of moment when N = 0. (a) Notation, (b) applied moment and (c) internal forces.

7.9 BIAXIAL BENDING

Columns with bending about two axes present a problem, because the interaction diagrams link N with M about only one axis. A simple and straightforward way to overcome this is to set one moment to zero and increase the moment about the other axis, as illustrated in Figure 7.16. The design then proceeds on the basis of the magnified moment alone. The first step is to identify which of the two moments should be eliminated. For a square column, this is usually the smallest moment, although if the column is wider about one axis than the other, then a quick check is needed.

Consider the column sketched in Figure 7.16. If

$$M_y/h' \geq M_z/b'$$

Then M_z is reduced to zero and M_y is replaced with

$$M_y' = M_y + M_z \times \frac{h'}{b'} \times \left(1 - \frac{N_{Ed}}{bhf_{ck}}\right) \qquad (7.55)$$

whereas if

$$M_y/h' < M_z/b'$$

then M_y is reduced to zero and

$$M_z' = M_z + M_y \times \frac{b'}{h'} \times \left(1 - \frac{N_{Ed}}{bhf_{ck}}\right) \qquad (7.56)$$

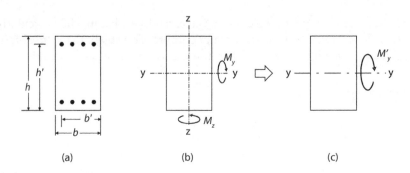

Figure 7.16 Analysis of biaxial bending. (a) Notation, (b) biaxial bending and (c) uniaxial bending.

Example 7.9: Construction of a *M–N* interaction diagram for a symmetric column

Figure 7.17 shows a cross section through a column. Plot the *M–N* interaction diagram and use it to determine if the column can resist a 280 kN.m moment, combined with a compression force of 1500 kN.

Basic data:
$A_{s1} = A_{s2} = 1963 \text{ mm}^2$
$f_{ck} = 40 \text{ N/mm}^2$

Point 1: Crushing

From Equation 7.45

$$C_c = 0.567 f_{ck} \times b \times h$$

$$C_c = 0.567 \times 40 \times 350 \times 350 \times 10^{-3} = 2778 \text{ kN}$$

And from Equation 7.46

$$N_{s1} = 0.87 A_{s1} f_{yk}$$

$$N_{s1} = N_{s2} = 0.87 \times 500 \times 1963 \times 10^{-3} = 854 \text{ kN}$$

And from Equation 7.43

$$N_{Rd} = C_c + N_{s1} + N_{s2}$$

$$N_{Rd} = 2778 + 854 + 854 = 4486 \text{ kN}$$

Since $N_{s1} = N_{s2}$, from Equation 7.44 $M_{Rd} = 0$ and Point 1 is located at the (0,4486) coordinates of the *M–N* interaction diagram.

Point 2: Balance

From Equation 7.52, the neutral axis depth is

$$x = 0.636d = 191 \text{ mm}$$

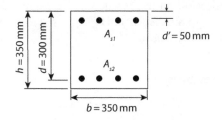

Figure 7.17 Column details.

From Equation 7.6

$$C_c = 0.567 f_{ck} b \times 0.8x$$

$$C_c = 0.567 \times 40 \times 350 \times 0.8 \times 191 \times 10^{-3} = 1213 \text{ kN}$$

From Equation 7.46

$$N_{s1} = -N_{s2} = 0.87 \times 500 \times 1963 \times 10^{-3} = 854 \text{ kN}$$

From force equilibrium (Equation 7.43)

$$N_{Rd} = C_c + N_{s1} + N_{s2} = 1213 + 854 - 854 = 1213 \text{ kN}$$

From Equations 7.48, 7.49 and 7.51, the lever arms are

$$z_{s1} = h/2 - d' = 0.350/2 - 0.05 = 0.125 \text{ m}$$

$$z_{s2} = -(d - h/2) = -(0.3 - 0.350/2) = -0.125 \text{ m}$$

$$z_c = h/2 - 0.4x = 0.35/2 - 0.4 \times 0.191 = 0.0986 \text{ m}$$

From moment equilibrium (Equation 7.50)

$$M_{Rd} = C_c z_c + N_{s1} z_{s1} + N_{s2} z_{s2}$$

$$M_{Rd} = 1213 \times 0.0986 + 854 \times 0.125 + (-854) \times (-0.125) = 333 \text{ kN.m}$$

Thus, the coordinates of Point 2 is (333,1213).

Point 3: Pure moment

From moment equilibrium (Equation 7.54)

$$M_{Rd} = -N_{s2}(d - d') = -(-854) \times (0.300 - 0.050) = 213 \text{ kN.m}$$

Thus, the coordinates of Point 3 is (213,0) and Figure 7.18 shows all the coordinates are plotted graphically to form an interaction diagram. This shows that the column can resist a 280 kN.m moment in addition to a 1500 kN compression force, since the (280,1500) coordinates lie within the design envelope. This graph is symmetrical about the vertical axis because it is a symmetrical cross section.

Example 7.10: Biaxial bending

Using the design chart shown in Figure 7.18, determine if the column shown in Figure 7.17 can resist 79 kN.m about the horizontal axis as sketched in Figure 7.17 and 35 kNm about the vertical axis, in addition to a compression force of 1500 kN.

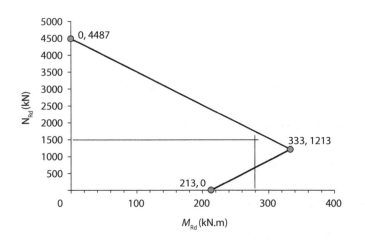

Figure 7.18 M–N interaction diagram.

In this case, $M_y/b' \geq M_z/b'$; therefore, M_z is reduced to zero and from Equation 7.55

$$M_y' = M_y + M_z \times \frac{b'}{b'} \times \left(1 - \frac{N_{Ed}}{bhf_{ck}}\right)$$

$$M_y' = 79 + 35 \times \frac{300}{300} \times \left(1 - \frac{1500 \times 10^3}{350 \times 350 \times 40}\right) = 103 \text{ kN.m}$$

Inspection of Figure 7.18 shows that the (103,1500) coordinates lie well within the $M-N$ diagram limits.

Example 7.11: Asymmetric column

Figure 7.19 shows a cross section through a column that is subjected to an axial compression force centred over the centre line and a uniaxial (+ve) moment as shown. Plot the $M-N$ interaction diagram, if $f_{ck} = 25 \text{ N/mm}^2$.

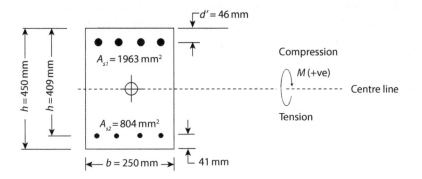

Figure 7.19 Cross section through a column with asymmetric reinforcement.

Point 1: Crushing

From Equations 7.45 through 7.47, the compression forces in the concrete and upper and lower rebar are

$$C_c = 0.567 \times 25 \times 250 \times 450 \times 10^{-3} = 1595 \text{ kN}$$

$$N_{s1} = 0.87 \times 500 \times 1963 \times 10^{-3} = 854 \text{ kN}$$

$$N_{s2} = 0.87 \times 500 \times 804 \times 10^{-3} = 350 \text{ kN}$$

And from force equilibrium (Equation 7.43)

$$N_{Rd} = 1595 + 854 + 350 = 2799 \text{ kN}$$

From Equations 7.48 and 7.49, the lever arms for the upper and lower rebar are

$$z_{s1} = h/2 - d' = 0.450/2 - 0.046 = 0.179 \text{ m}$$

$$z_{s2} = -(d - h/2) = -(0.409 - 0.450/2) = -0.184 \text{ m}$$

From the moment equilibrium (Equation 7.44)

$$M_{Rd} = N_{s1}z_{s1} + N_{s2}z_{s2} = 854 \times 0.179 + 350 \times (-0.184) = 88 \text{ kN.m}$$

Therefore, the coordinates of Point 1 are (88,2799).

Point 2: Balanced failure (+ve moments)

From Equation 7.52, the neutral axis depth (x) = 260 mm and from Equation 7.6

$$C_c = 0.567 \times f_{ck} \times b \times 0.8x$$

$$C_c = 0.567 \times 25 \times 250 \times 0.8 \times 260 \times 10^{-3} = 737 \text{ kN}$$

The strain in the compression steel is greater than the yield strain (0.002) when calculated using Equation 7.53; therefore, Equation 7.46 is used to calculate N_{s1}. From the force equilibrium

$$N_{Rd} = C_c + N_{s1} + N_{s2}$$

$$N_{Rd} = 737 + 854 - 350 = 1241 \text{ kN}$$

And from Equation 7.51

$$z_c = h/2 - 0.4x$$

$$z_c = 0.450/2 - 0.4 \times 0.260 = 0.121 \text{ m}$$

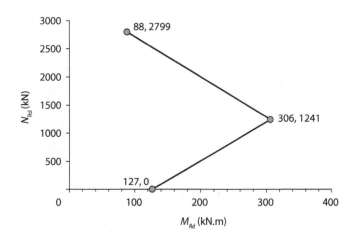

Figure 7.20 M–N interaction diagram for an asymmetric column.

From the moment equilibrium (Equation 7.50)

$$M_{Rd} = C_c z_c + N_{s1} z_{s1} + N_{s2} z_{s2}$$

$$M_{Rd} = 737 \times 0.121 + 854 \times 0.179 + (-350) \times (-0.184) = 306 \text{ kN.m}$$

Therefore, the coordinates of Point 2 are (306, 1241).

Point 3: Pure (+ve) moment

The tensile force in the lower rebar is balanced by an equal and opposite force in the upper rebar; therefore, from Equation 7.54

$$M_{Rd} = -N_{s2}(d - d')$$

$$M_{Rd} = -(-350) \times (0.409 - 0.046) = 127 \text{ kN.m}$$

The coordinates of Point 3 are (127,0) and all three coordinates are plotted in the interaction diagram shown in Figure 7.20.

Example 7.12: Asymmetric column

Construct the M–N diagram for the column considered in Example 7.11 including the negative moment region (the coordinates of Points 1, 2 and 3 are shown in Figure 7.20).

Point 4: Pure bending (–ve moment)

Under negative moments, the top steel is in tension and the bottom in compression; therefore, from Equation 7.46 and 7.47

$$N_{s1} = -854 \text{ kN}$$

$$N_{s2} = +350 \text{ kN}$$

and from force equilibrium ($N_{Rd} = 0$)

$$C_c + N_{s1} + N_{s2} = 0$$

$$C_c = +854 - 350 = 504 \text{ kN}$$

From Equation 7.6

$$C_c = 0.567 \times f_{ck} \times b \times 0.8x$$

$$504 \times 10^3 = 0.567 \times 25 \times 250 \times 0.8x$$

$$x = 178 \text{ mm}$$

The lever arm for the concrete stress block is negative because C_c is below the neutral axis; therefore, from Equation 7.51

$$z_c = -(h/2 - 0.4x)$$

$$z_c = -(0.450/2 - 0.4 \times 0.178) = -0.154 \text{ m}$$

From moment equilibrium (Equation 7.50)

$$M_{Rd} = C_c z_c + N_{s1} z_{s1} + N_{s2} z_{s2}$$

$$M_{Rd} = 504 \times (-0.154) - 854 \times 0.179 + 350 \times (-0.184) = -295 \text{ kN.m}$$

Therefore, the coordinates of Point 5 are (−295, 0).

Point 5: Balanced failure (−ve moments)

From Equation 7.52, the neutral axis depth, when measured from the bottom of the section, is

$$x = 0.636 \times (450 - 46) = 257 \text{ mm}$$

From Equation 7.6

$$C_c = 0.567 \times 25 \times 250 \times 0.8 \times 257 \times 10^{-3} = 729 \text{ kN}$$

Under negative moments, the top steel is in tension and the bottom in compression; therefore, from Equations 7.46 and 7.47

$$N_{s1} = -854 \text{ kN}$$

$$N_{s2} = +350 \text{ kN}$$

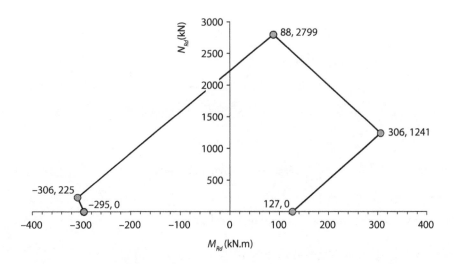

Figure 7.21 M–N interaction diagram for the asymmetric column shown in Figure 7.19.

And from force equilibrium (Equation 7.43)

$$N_{Rd} = C_c + N_{s1} + N_{s2}$$

$$N_{Rd} = 729 - 854 + 350 = 225 \text{ kN}$$

The lever arm for the concrete is negative, because C_c is below the neutral axis; therefore, from Equation 7.51

$$z_c = -(h/2 - 0.4x)$$

$$z_c = -(0.450/2 - 0.4 \times 0.257) = -0.122 \text{ m}$$

From moment equilibrium (Equation 7.50)

$$M_{Rd} = C_c z_c + N_{s1} z_{s1} + N_{s2} z_{s2}$$

$$M_{Rd} = 729 \times (-0.122) - 854 \times 0.179 + 350 \times (-0.184) = -306 \text{ kN.m}$$

Therefore, the coordinates of Point 5 are (−306, 225).

Example 7.13: Non-rectangular, symmetrical column

Plot the interaction diagram for the column shown in Figure 7.22 if $f_{yk} = 40 \text{ N/mm}^2$ and each row of rebar has an area of 630 mm².

Point 1: Crushing

From Equation 7.45

$$C_c = 0.567 \times 40 \times 400 \times 250 \times 10^{-3} = 2268 \text{ kN}$$

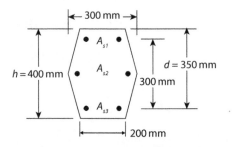

Figure 7.22 Column details.

And the compression strength of each row of rebar is

$$N_{s1} = N_{s2} = N_{s3} = 0.87 \times 500 \times 630 \times 10^{-3} = 274 \text{ kN}$$

And from Equation 7.43

$$N_{Rd} = 2268 + 3 \times 274 = 3090 \text{ kN}$$

Since the column is symmetrical, crushing occurs at zero moment and Point 1 is therefore located at the (0,3090) coordinates of the interaction diagram.

Point 2: Balance

From Equation 7.52, the neutral axis depth is

$$x = 0.636 \times 350 = 223 \text{ mm}$$

The compression force in the concrete stress block must now be determined. The area of a trapezoid (see Figure 7.23) is

$$A = \frac{g}{2}(e+f) \tag{7.57}$$

And the distance from the top edge to the centroid is

$$\bar{y} = \frac{g}{3} \times \frac{2e+f}{e+f} \tag{7.58}$$

The depth of the stress block from the top edge of the column is $0.8x = 178$ mm. From Equation 7.57, the area of the compression stress block is $A = 43,637$ mm² and the corresponding compression force is

$$C_c = 0.567 f_{ck} A$$

$$C_c = 0.567 \times 40 \times 43637 \times 10^{-3} = 990 \text{ kN}$$

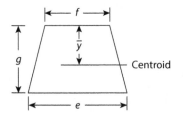

Figure 7.23 General trapezoid.

The force in the top and bottom rows of rebar is

$$N_{s1} = -N_{s3} = 274 \text{ kN}$$

Using similar triangles from the strain distribution, the strain in the middle row of rebar (ε_{s2}) is

$$\frac{x}{0.0035} = \frac{x - h/2}{\varepsilon_{s2}}$$

$$\varepsilon_{s2} = \frac{0.0035 \times (223 - 200)}{223} = 360 \times 10^{-6}$$

And the corresponding stress is

$$\sigma_{s2} = \varepsilon_{s2} E_s$$

$$\sigma_{s2} = 360 \times 10^{-6} \times 210000 = 76 \text{ N/mm}^2$$

which develops the following force

$$N_{s2} = 0.87 \sigma_s A_{s2} = 0.87 \times 76 \times 630 \times 10^{-3} = 42 \text{ kN}$$

From the force equilibrium

$$N_{Rd} = C_c + N_{s1} + N_{s2} + N_{s3}$$

$$N_{Rd} = 990 + 42 + 274 - 274 = 1032 \text{ kN}$$

The lever arm distances for the rows of rebar are

$$z_{s1} = 0.150 \text{ m}$$

$$z_{s2} = 0 \text{ m}$$

$$z_{s3} = -0.150 \text{ m}$$

From Equation 7.58, $\bar{y} = 94.6$ mm; therefore,

$$z_c = h/2 - \bar{y} = 0.400/2 - 0.0946 = 0.1054 \text{ m}$$

From the moment equilibrium

$$M_{\text{Rd}} = C_c z_c + N_{s1} z_{s1} + N_{s2} z_{s2} + N_{s3} z_{s3}$$

$$M_{\text{Rd}} = 990 \times 0.1054 + 274 \times 0.150 + 42 \times 0 + (-274) \times (-0.150) = 187 \text{ kN.m}$$

Point 3: Pure moment

From horizontal equilibrium (for simplicity, ignoring N_{s1})

$$C_c + N_{s2} + N_{s3} = 0$$

$$C_c = 2 \times 274 = 548 \text{ kN}$$

since

$$C_c = 0.567 f_{ck} A$$

Rearranging to give the area of the concrete stress block

$$A = \frac{548 \times 10^3}{0.567 \times 40} = 24162 \text{ mm}^2$$

which has a depth (g) of 106.6 mm, solved as a quadratic equation from Equation 7.57, with a centroid at $\bar{y} = 55.3$ mm. Taking the moment about the centroid of the concrete block

$$M_{\text{Rd}} = 274 \times (0.35 - 0.0553) + 274 \times (0.2 - 0.0553) = 120 \text{ kN.m}$$

Figure 7.24 shows Points 1, 2 and 3 plotted graphically to form the design chart.

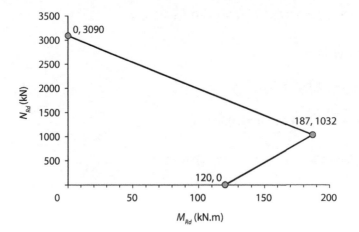

Figure 7.24 M–N interaction diagram.

Problems

Solutions to these problems are provided at https://www.crcpress.com/9781498741217

P.7.1. A 375 mm wide and 600 mm deep RC beam spans 8.5 m between simple supports. It supports its self-weight in addition to a uniformly distributed unfactored imposed load of 21 kN/m. The strengths are f_{yk} = 500 N/mm² for the reinforcement and f_{ck} = 35 N/mm² for the concrete. A cover of 25 mm is provided and the density of concrete = 25 kN/m³.
   a. Determine the ULS moment and shear force.
   b. Assuming 8 mm diameter shear links and 40 mm diameter main reinforcement, determine the effective depth used to begin the strength calculations.
   c. Determine the minimum area of tension steel.
   d. Determine the distance from the supports to the 50% midspan moment point.
   e. If 50% of the main bending reinforcement is to be curtailed, determine the distance from the centre of the end supports to the curtailment point. Assume 25 mm rebar.
   Ans. (a) M_{Ed} = 353 kN.m, V_{Ed} = 166 kN, (b) 547 mm, (c) 1626 mm², (d) 1250 mm and (e) 425 mm.

P.7.2. A 400 mm wide and 350 mm deep RC beam is reinforced with 8 mm diameter shear links, four 32 mm diameter tension rebars and four 20 mm compression rebars. The cover provided is 25 mm, f_{ck} = 40 N/mm² and f_{yk} = 500 N/mm².
   a. Determine the ULS moment capacity.
   b. The beam spans 7 m between simple supports and is subjected to a ULS UDL of 58 kN/m (inclusive of self-weight). Determine the distance from the support to the 50% moment point.
   c. Determine the distance from the end of the beam that 50% of the tension reinforcement can be curtailed.
   Ans. (a) T = 1399 kN, C_s = 546.5 kN, C_c = 852 kN, x = 117.5 mm, M = 357.5 kN.m, (b) 1025 mm and (c) 65 mm.

P.7.3. A 375 mm wide and 500 mm deep RC beam spans 9 m between simple supports. It supports its self-weight plus a 10 kN/m unfactored uniformly distributed imposed load (UDL) plus a centrally applied unfactored point load comprising 30 kN dead load and 35 kN imposed load. The characteristic strengths are f_{yk} = 500 N/mm² for the reinforcement and f_{ck} = 35 N/mm² for the concrete. A cover of 25 mm is provided.
   a. Determine the ULS design moment and shear force.
   b. Assuming 8 mm diameter shear links and 40 mm diameter main rebar, determine the effective depth used to begin the strength calculations.
   c. Determine the minimum area of tension reinforcement required.
   d. It has been decided to use 8 mm diameter shear links. Determine the minimum spacing at the ends of the beam.
   e. Determine the distance from the supports that the shear link spacing can be increased to 200 mm.
   Ans. (a) M_{Ed} = 424.9 kN.m, V_{Ed} = 142.35 kN, (b) 447 mm, (c) 2641 mm², (d) 123.7 mm and (e) 2552 mm.

P.7.4. Figure 7.25 shows a cross section through a column.

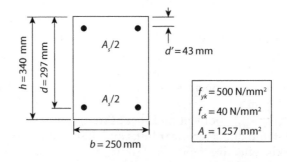

Figure 7.25 Cross section through a column.

a. Determine the crushing strength (i.e. Point 1 on Figure 7.12).
b. Determine the moment and axial force at 'balanced failure', i.e., Point 2 on Figure 7.12.
c. Determine the bending strength in the absence of compression (i.e., Point 3).
Ans. (a) 2475 kN, (b) 150 kN.m, 857 kN and (c) 69 kN.m.

Chapter 8

Prestressed structures

The purpose of Prestressed Concrete (PSC) is to minimise tensile stresses by inducing compression. This is achieved by the tensioning of high strength steel strands using hydraulic jacks. The strands usually comprise seven galvanised steel wires twisted together to form a rope. This overcomes concrete's weakness in tension and helps control cracking, increases stiffness, reduces deflections, reduces material costs, as well as allowing for the design of elegant and structures.

There are two main types: pre-tensioning and post-tensioning. Pre-tensioning involves casting the concrete around tensioned tendons. This is common for manufacturing precast concrete members, such as railway sleepers and bridge beams (see Figure 8.1). This chapter concentrates mainly on post-tensioned concrete, where the tendons are stressed after the concrete has hardened. Post-tensioning is often used for curved structures, such as that illustrated in Figure 8.2a. The tendon forces will not cause buckling, as would be the case for a curved member subjected to an externally applied compression force (see Figure 8.2b). This makes the technique popular for constructing tanks and silos, since the prestress can counterbalance the stresses induced by the liquids and thus maintain the concrete in a state of compression. This reduces cracking and leakages, which can be particularly useful when storing hazardous liquids.

Prestressing strands can be bonded to the concrete using grout injected into the ducts, or they can be left unbonded. This chapter assumes the strands are bonded, and it also assumes that the concrete section remains uncracked when supporting SLS loads. This is known as *full prestress*. Experts in prestressing sometimes allow the section to crack, which is a state

Figure 8.1 Underside of a bridge constructed using precast and pre-tensioned concrete beams.

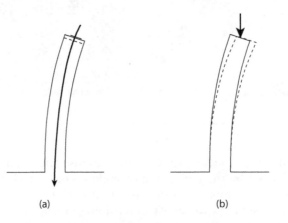

(a)   (b)

Figure 8.2 Difference in response due to 'internal' prestressing force and 'externally' applied load. (a) Internal prestressing force produces no sideways movement (if applied through centroid) and (b) external load results in sideways sway.

known as 'partial prestress'. This requires a spreadsheet solution and is beyond the scope of this book, which aims to provide an understanding of the basic design principles for undergraduates.

8.1 INTRODUCTION TO THE BASIC THEORY

If a cross section is subjected to a prestressing force located on the centroid (Figure 8.3a), the stress is

$$\sigma_T = \sigma_B = \frac{P}{A}$$

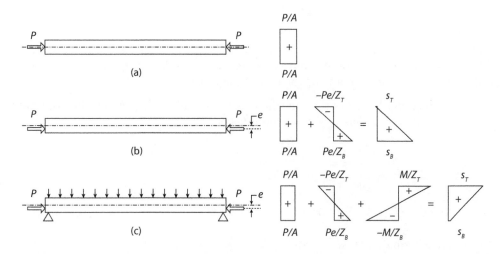

Figure 8.3 Illustration of the stresses induced due to prestressing (compression is +ve). (a) Tendon on centroid, NO gravity moment, (b) tendon below centroid, NO gravity moment and (c) tendon below centroid + gravity moment.

where
σ_T is the stress in the top of the beam.
σ_B is the stress in the bottom of the beam.
P is the initial prestress force.
A is the cross-sectional area.

If the tendon is repositioned to below the neutral axis as illustrated in Figure 8.3b, then

$$\sigma_T = \frac{P}{A} - \frac{Pe}{Z_T}$$

$$\sigma_B = \frac{P}{A} + \frac{Pe}{Z_B}$$

where
e is the distance from the prestressing force to the neutral axis, known as the *eccentricity*.
Z_T and Z_B are the elastic section moduli required to give the bending stresses at the top and bottom fibres of the section, respectively.

If a moment due to gravity loads is added (Figure 8.3c), then

$$\sigma_T = \frac{P}{A} - \frac{Pe}{Z_T} + \frac{M}{Z_T} \tag{8.1}$$

$$\sigma_B = \frac{P}{A} + \frac{Pe}{Z_B} - \frac{M}{Z_B} \tag{8.2}$$

Example 8.1: Beam with straight tendon

Figure 8.4 shows a 400 mm wide and, 800 mm deep beam that spans 14 m between simple supports. It supports an imposed load of 21 kN/m and is prestressed with a force of 1500 kN. If the tendon is straight and is located 260 mm below the centroid, sketch the stress distribution at midspan and at the end of the beam.

Since this is a rectangular cross section, the elastic moduli of the top and bottom fibres of the section are equal. The elastic section modulus is

$$Z = \frac{I}{y} = \frac{bh^3}{12} \times \frac{2}{h} = \frac{bh^2}{6}$$

(8.3)

$Z = 400 \times 800^2/6 = 42.7 \times 10^6$ mm^3

And the cross-sectional area is

$A = 800 \times 400 = 320 \times 10^3$ mm^2

From Equation 8.1, the stress at the ends of the beam, where $M = 0$, is

$$\sigma_T = \frac{P}{A} - \frac{Pe}{Z_T} + \frac{M}{Z_T}$$

$$\sigma_T = \frac{1500 \times 10^3}{320 \times 10^3} - \frac{1500 \times 10^3 \times 260}{42.7 \times 10^6} + 0$$

$$\sigma_T = 4.7 - 9.1 + 0 = -4.4 \text{ N/mm}^2$$

And from Equation 8.2

$$\sigma_B = 4.7 + 9.1 - 0 = +13.8 \text{ N/mm}^2$$

The uniformly distributed Serviceability Limit State (SLS) load (assuming concrete density is 25.0 kN/m^3) is

$w = 25 \times 0.4 \times 0.8 + 21 = 29$ kN/m

And the midspan moment is

$M = 29 \times 14^2/8 = 710.5$ kN.m

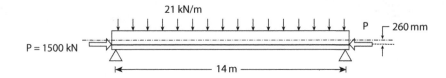

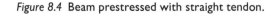

Figure 8.4 Beam prestressed with straight tendon.

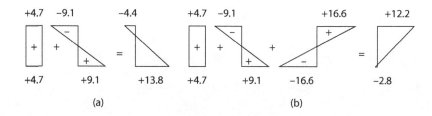

Figure 8.5 Stress distributions in N/mm² for Example 8.1. (a) Stress distribution at support and (b) stress distribution at midspan.

From Equations 8.1 and 8.2, the stresses at midspan (top and bottom) are

$$\sigma_T = \frac{P}{A} - \frac{Pe}{Z_T} + \frac{M}{Z_T}$$

$$\sigma_T = \frac{1500 \times 10^3}{320 \times 10^3} - \frac{1500 \times 10^3 \times 260}{42.7 \times 10^6} + \frac{710.5 \times 10^6}{42.7 \times 10^6}$$

$$\sigma_T = 4.7 - 9.1 + 16.6 = +12.2 \text{ N/mm}^2$$

$$\sigma_B = 4.7 + 9.1 - 16.6 = -2.8 \text{ N/mm}^2$$

These stresses are shown graphically in Figure 8.5. This demonstrates that the worst tensile and compression stresses occur at the supports, rather than midspan. This problem can be easily overcome by using a draped tendon, as demonstrated in Example 8.2.

Example 8.2: Beam with draped tendon

A 400 mm wide and 800 mm deep simply supported beam spans 14 m (see Figure 8.6). It supports an imposed load of 45 kN/m. The tendon is located on the centroid at the supports and 375 mm below the centroid at midspan. Sketch the stress distribution at midspan and at the ends of the beam.

From Equations 8.1 and 8.2, the stresses at the support's top and bottom are

$$\sigma_T = \frac{P}{A} - \frac{Pe}{Z_T} + \frac{M}{Z_T}$$

$$\sigma_T = \frac{2500 \times 10^3}{320 \times 10^3} - \frac{2500 \times 10^3 \times 0}{42.7 \times 10^6} + \frac{0}{42.7 \times 10^6}$$

Figure 8.6 Beam with draped tendon.

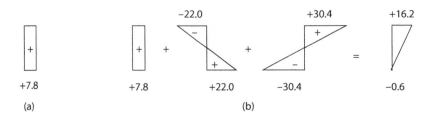

Figure 8.7 Stress distributions in N/mm² for Example 8.2. (a) Support stresses and (b) midspan stresses.

$$\sigma_T = +7.8 - 0 + 0 = +7.8 \text{ N/mm}^2$$

$$\sigma_B = +7.8 + 0 - 0 = +7.8 \text{ N/mm}^2$$

The UDL (w) and midspan moment (M) are

$$w = 25 \times 0.4 \times 0.8 + 45 = 53 \text{ kN/m}$$

$$M = 53 \times 14^2/8 = 1298.5 \text{ kN.m}$$

From Equations 8.1 and 8.2, the stresses at midspan (top and bottom) are

$$\sigma_T = \frac{P}{A} - \frac{Pe}{Z_T} + \frac{M}{Z_T}$$

$$\sigma_T = 7.8 - \frac{2500 \times 10^3 \times 375}{42.7 \times 10^6} + \frac{1298.5 \times 10^6}{42.7 \times 10^6}$$

$$\sigma_T = 7.8 - 22.0 + 30.4 = +16.2 \text{ N/mm}^2$$

$$\sigma_B = +7.8 + 22.0 - 30.4 = -0.6 \text{ N/mm}^2$$

Figure 8.7 shows the stress at the support and midspan. When these are compared with those from Example 8.1 it can be seen that the stresses are more favourable, even though the load increased from 21 kN/m to 45 kN/m. This explains why draped tendons are used if possible.

8.2 SLS DESIGN

The prime objective during the SLS design is to prevent either tensile stresses causing cracking or compressive stresses causing crushing. Failure can occur immediately after loading the tendons, known as *at transfer*, or under the full SLS (working) loads.

At transfer, the gravity loads will be lower than the SLS loads and the tendon forces may cause the beam to arch, leading to possible cracking at the top and crushing of the concrete at the bottom (see Figure 8.8a).

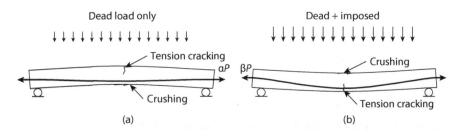

Figure 8.8 Illustration of conditions at transfer and under SLS loading. (a) Loading 'at transfer' and (b) SLS loading.

Under full SLS loading, these failure modes are reversed, with the top of the beam potentially failing by crushing and the bottom by cracking (see Figure 8.8b). Another complication is the difference in concrete strength at transfer and under SLS conditions. The fresh concrete may be low strength at transfer, whereas it would be full strength under the SLS conditions. In addition, the loss of prestress under SLS conditions will be greater than at transfer. Using these factors, four design inequalities are formed.

At transfer,

$$\text{Tendon force} = \alpha \times P$$

And under SLS loads

$$\text{Tendon force} = \beta \times P$$

where
  α is the short-term loss factor, accounting for friction, elastic shortening, and anchorage draw-in.
  β is the total loss factor, which accounts for short-term losses + long-term losses from shrinkage, creep, and relaxation of stress in the steel.
  P is the jacking force applied to the prestress stands.

When the tendons are first stressed, the main load is from the prestress, which may cause the top of the section to crack in tension and the bottom to be liable to crushing (see Figure 8.8a). The stress at the top of the section must be less than or equal to the cracking stress, i.e.,

$$\frac{\alpha P}{A} - \frac{\alpha P e}{Z_T} + \frac{M_{dl}}{Z_T} \geq \sigma_{\min.t} \qquad (8.4)$$

And at the bottom of the section, crushing is the failure mode and

$$\frac{\alpha P}{A} + \frac{\alpha P e}{Z_B} - \frac{M_{dl}}{Z_B} \leq \sigma_{\max.t} \qquad (8.5)$$

where
  $\sigma_{\min.t}$ is the minimum permissible (tensile) stress at transfer.
  $\sigma_{\max.t}$ is the maximum permissible compression stress at transfer.
  M_{dl} is the moment at transfer, which is usually just the dead load moment.

Under full SLS dead and imposed loading (Figure 8.8b), the loading increases and the concrete gains its full strength, although the prestress force diminishes further due to long-term prestress losses. At the top of the section, crushing becomes the failure mode and

$$\frac{\beta P}{A} - \frac{\beta Pe}{Z_T} + \frac{M_{\text{sls}}}{Z_T} \leq \sigma_{\text{max.sls}} \tag{8.6}$$

And at the bottom, tension cracking needs to be prevented and

$$\frac{\beta P}{A} + \frac{\beta Pe}{Z_B} - \frac{M_{\text{sls}}}{Z_B} \geq \sigma_{\text{min.sls}} \tag{8.7}$$

where
  $\sigma_{\text{max.sls}}$ is the maximum permissible compression stress.
  $\sigma_{\text{min.sls}}$ is the minimum permissible (tension) stress.
  M_{sls} is the SLS moment.

8.2.1 Member sizing

Equations 8.4 through 8.7 can be reconfigured to provide minimum values for section moduli of the member

$$Z_T \geq \frac{\alpha M_{\text{sls}} - \beta M_{\text{dl}}}{\alpha \sigma_{\text{max.sls}} - \beta \sigma_{\text{min}.t}} \tag{8.8}$$

$$Z_B \geq \frac{\alpha M_{\text{sls}} - \beta M_{\text{dl}}}{\beta \sigma_{\text{max}.t} - \alpha \sigma_{\text{min.sls}}} \tag{8.9}$$

8.2.2 The permissible ranges of tendon force

The four inequalities (Equations 8.4 through 8.7) can be rearranged to provide four inequalities defining the acceptable limits of the tendon force. Considering the first inequality (Equation 8.4)

$$P\left(\frac{\alpha}{A} - \frac{\alpha e}{Z_T}\right) \geq \sigma_{\text{min}.t} - \frac{M_{\text{dl}}}{Z_T} \tag{8.10}$$

And Equation 8.5 through 8.7 each rearrange to

$$P\left(\frac{\alpha}{A} + \frac{\alpha e}{Z_B}\right) \leq \sigma_{\text{max}.t} + \frac{M_{\text{dl}}}{Z_B} \tag{8.11}$$

$$P\left(\frac{\beta}{A} - \frac{\beta e}{Z_T}\right) \leq \sigma_{\text{max.sls}} - \frac{M_{\text{sls}}}{Z_T} \tag{8.12}$$

$$P\left(\frac{\beta}{A} + \frac{\beta e}{Z_B}\right) \geq \sigma_{\text{min.sls}} + \frac{M_{\text{sls}}}{Z_B} \tag{8.13}$$

Remember that the sign of an inequality changes when divided by a negative number. For example:

$$P \times (-2) < -8$$

Dividing both sides by –2 reverses the sign, i.e.,

$$P > \frac{-8}{-2}$$

The negatives cancel; therefore and

$$P > 4$$

Example 8.3: Beam sizing and prestressing force calculation

A 500 mm wide rectangular beam spans 12 m between simple supports and supports a 5 kN/m imposed load. The limiting stresses and prestress loss factors are listed in the basic data below.

1. Determine the minimum beam depth required to support the load.
2. If the depth of the beam is set at 0.35 m and the tendons are located 0.125 m below the centroid at midspan, determine the minimum and maximum values of tendon force.

Basic data
Loss factors: $\alpha = 0.92$ and $\beta = 0.82$
At transfer: $\sigma_{min} = -1.0$ N/mm^2 (tension) and $\sigma_{max} = 18$ N/mm^2
At SLS: $\sigma_{min} = 0.0$ N/mm^2 and $\sigma_{max} = 20$ N/mm^2

1. If the concrete density is 25 kN/m^2, then the dead load of a 0.5 m wide and h deep beam is

$$w_{dl} = 25 \times 0.5 \times h = 12.5h \text{ kN/m}$$

The dead load and SLS load midspan moments are

$$M_{dl} = \frac{12.5h \times 12^2}{8} = 225h \text{ kN.m} \tag{8.14}$$

$$M_{sls} = M_{dl} + \frac{5 \times 12^2}{8} = (225h + 90) \text{ kN.m} \tag{8.15}$$

From Equation 8.3, the section modulus in m^3 for a 0.5 m wide beam of depth h is

$$Z = \frac{0.5 \times h^2}{6} = \frac{h^2}{12}$$

Looking at the first inequality (Equation 8.8)

$$Z_T \geq \frac{\alpha M_{sls} - \beta M_{dl}}{\alpha \sigma_{max.sls} - \beta \sigma_{min.t}}$$

It is convenient to carry out Equation 8.8 calculations in units of kN and metres. This only works if the units of stress are converted from N/mm² to kN/m². In this example, 20 N/mm² = 20 × 10³ kN/m² and inputting into Equation 8.8

$$\frac{h^2}{12} \geq \frac{0.92\,(225h+90)-0.82\times225h}{0.92\times20\times10^3-0.82\times(-1.0)\times10^3}$$

which simplifies to

$$1601h^2-22.5h-82.8\geq0$$

The roots of this quadratic equation are 0.235 m and −0.220 m. Repeating for the second inequality (Equation 8.9)

$$\frac{h^2}{12} \geq \frac{0.92\,(225h+90)-0.82\times225h}{0.82\times18\times10^3-0.92\times0.0\times10^3}$$

Simplifying

$$1230h^2-22.5h-82.8\geq0$$

The roots are 0.269 m and −0.250 m. Thus, the minimum depth is the largest of the four roots, i.e., h ≥0.269 m.

2. Inputting a depth of 0.35 m into the previous working shown in Equations 8.14 and 8.15, the dead load and SLS moments are

$$M_{dl}=225\times0.35=78.75\text{ kN.m}$$

$$M_{sls}=225\times0.35+90=168.75\text{ kN.m}$$

The cross-sectional area and elastic section modulus are

$$A=0.5\times0.35=0.175\text{ m}^2$$

$$Z=\frac{bh^2}{6}=\frac{0.5\times0.35^2}{6}=0.01021\text{ m}^3$$

Inputting these into the inequalities provides the limits on P, from Equation 8.10

$$P\left(\frac{\alpha}{A}-\frac{\alpha e}{Z_T}\right)\geq\sigma_{\min.t}-\frac{M_{dl}}{Z_T}$$

$$P\left(\frac{0.92}{0.175}-\frac{0.92\times0.125}{0.01021}\right)\geq-1\times10^3-\frac{78.75}{0.01021}$$

$$-6.006\,P\geq-8713$$

Dividing though by −6.006 changes the sign

$$P \leq \frac{-8713}{-6.006}$$

The negatives cancel; therefore,

$$P \leq 1450 \text{ kN}$$

The second inequality (Equation 8.11)

$$P\left(\frac{0.92}{0.175} + \frac{0.92 \times 0.125}{0.01021}\right) \leq 18 \times 10^3 + \frac{78.75}{0.01021}$$

$$P \leq 1556 \text{ kN}$$

The third inequality (Equation 8.12)

$$P\left(\frac{0.82}{0.175} - \frac{0.82 \times 0.125}{0.01021}\right) \leq 20 \times 10^3 - \frac{168.75}{0.01021}$$

$$-5.353\,P \leq 3472$$

Dividing though by −5.353 changes the sign

$$P \geq \frac{3472}{-5.353}$$

$$P \geq -648 \text{ kN}$$

And the final inequality (Equation 8.13)

$$P\left(\frac{0.82}{0.175} + \frac{0.82 \times 0.125}{0.01021}\right) \geq 0 + \frac{168.75}{0.01021}$$

$$P \geq 1122 \text{ kN}$$

The four inequalities, $P \leq 1450$ kN, $P \leq 1556$ kN, $P \geq -648$ kN, and $P \geq 1122$ kN, can be satisfied within the following limits:

$$1122 \text{ kN} \leq P \leq 1450 \text{ kN}$$

8.2.3 Determining the allowable tolerance in the positioning of the prestressing tendons

Civil engineering contractors require information on the accuracy needed in the positioning of the prestressing strands. The tolerance in cable zone positioning is illustrated in Figure 8.9, and the tolerance can be determined by manipulating the previously derived inequalities.

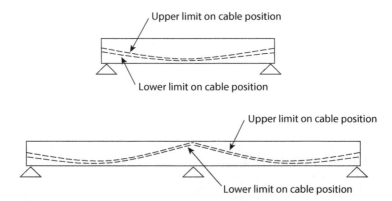

Figure 8.9 Tendon positioning tolerance.

The inequalities (Equations 8.10 through 8.13) rearrange to

$$e \leq \frac{M_{dl} - \sigma_{min.t}Z_T}{\alpha P} + \frac{Z_T}{A} \tag{8.16}$$

$$e \leq \frac{M_{dl} + \sigma_{max.t}Z_B}{\alpha P} - \frac{Z_B}{A} \tag{8.17}$$

$$e \geq \frac{M_{sls} - \sigma_{max.sls}Z_T}{\beta P} + \frac{Z_T}{A} \tag{8.18}$$

$$e \geq \frac{M_{sls} + \sigma_{min.sls}Z_B}{\beta P} - \frac{Z_B}{A} \tag{8.19}$$

By using these, it is possible to determine the region within the beam in which the tendons can be placed without overstressing the concrete. It should be noted that e is positive when the tendon is below the centroid.

Example 8.4: Tolerance in tendon position

A bridge spans 25 m and comprises 600 mm deep precast concrete beams. Each is 1000 mm wide and is subjected to an imposed load of 12 kN/m. The maximum compressive stress in the concrete is limited to 22 N/mm² (or 22,000 kN/m²) and no tensile stresses are allowed to develop. A total of 8% of the prestress is lost at transfer and 18% during serviceability. Each 1000 mm wide unit is prestressed with a force of 8000 kN. Determine the tolerance for the cable zone at the ends of the beam and at midspan.

The prestress loss factors at transfer (α) and under SLS loads (β) are

$$\alpha = 1.0 - \frac{8}{100} = 0.92 \left(8\% \text{ loss at transfer}\right)$$

$$\beta = 1.0 - \frac{18}{100} = 0.82 \left(18\% \text{ loss at SLS}\right)$$

The elastic section modulus (Z) from Equation 8.3 and cross-sectional area (A) are

$$Z = \frac{bh^2}{6} = \frac{1.0 \times 0.6^2}{6} = 0.06 \text{ m}^2$$

$$A = 0.60 \times 1.0 = 0.6 \text{ m}^2$$

At transfer (i.e. when the tendons are stressed), the UDL and midspan moments are

$$w_{dl} = 25 \times 1.0 \times 0.6 = 15 \text{ kN/m}$$

$$M_{dl} = 15 \times 25^2/8 = 1172 \text{ kN.m}$$

And under full SLS loading

$$w_{sls} = 15 + 12 = 27 \text{ kN/m}$$

$$M_{sls} = 27 \times 25^2/8 = 2109 \text{ kN.m}$$

Solving the inequality in Equation 8.16:

$$e \le \frac{M_{dl} - \sigma_{min.t} Z_T}{\alpha P} + \frac{Z_T}{A}$$

$$e \le \frac{M_{dl} - 0 \times 10^3 \times 0.06}{0.92 \times 8000} + \frac{0.06}{0.6}$$

$$e \le 0.000136 M_{dl} + 0.1$$

At the supports, $M_{dl} = 0$ kN.m $e \le 0.1$ m

At midspan, $M_{dl} = 1172$ kN.m $e \le 0.259$ m

Repeating for the second inequality (Equation 8.17):

$$e \le \frac{M_{dl} + \sigma_{max.t} Z_B}{\alpha P} - \frac{Z_B}{A}$$

$$e \le \frac{M_{dl} + 22 \times 10^3 \times 0.06}{0.92 \times 8000} - \frac{0.06}{0.6}$$

$$e \le 0.000136 M_{dl} + 0.0794$$

At the supports, $M_{dl} = 0$ kN.m $e \le 0.0794$ m

At midspan, $M_{dl} = 1172$ kN.m $e \le 0.239$ m

The third inequality (Equation 8.18):

$$e \geq \frac{M_{sls} - 22 \times 10^3 \times 0.06}{0.82 \times 8000} + \frac{0.06}{0.6}$$

$$e \geq 0.000152\, M_{sls} - 0.101$$

At the supports, $M_{sls} = 0$ kN.m $e \geq -0.101$ m

And at midspan, $M_{sls} = 2109$ kN.m $e \geq 0.220$ m

And the final inequality (Equation 8.19):

$$e \geq \frac{M_{sls} + 0.0 \times 10^3 \times 0.06}{0.82 \times 8000} - \frac{0.06}{0.6}$$

$$e \geq 0.000152\, M_{sls} - 0.1$$

Supports $e \geq -0.1$ m

Midspan $e \geq 0.221$ m

The final tolerance limits on cable position are

Supports -0.10 m $\leq e \leq 0.0794$ m

Midspan 0.221 m $\leq e \leq 0.239$ m

Inspection of these limits shows that there is a great deal of tolerance on cable position at the ends of the beam, although the cable zone width is only 18 mm at midspan.

8.2.4 Prestress losses

There are two types of prestress loss: *short-term* and *long-term losses*. Short-term losses occur immediately the tendons are stressed: whereas long-term losses occur during the working life. The short-term losses are accounted for by the α-*factor* applied to the prestress force and total losses by the β-*factor*, which include the sum of the short-term and long-term losses.

Short-term losses include the following:

1. Anchorage draw-in
2. Elastic shortening of the concrete
3. Friction during tensioning of tendons

Long-term losses include the following:

1. Concrete shrinkage
2. Relaxation of the steel tendons
3. Concrete creep

In this book each of these is considered separately when calculating the α and β factors, although that is a conservative approach.

8.2.4.1 Anchorage draw-in

The tendons are stressed using jacks, which when released draw wedges into anchor blocks. Some tendon force is lost as the wedges are drawn into the anchors, although this is often ignored because friction normally prevents the loss reaching a critical section. However, anchorage draw-in can be important with unbonded straight tendons and the loss is easily calculated.

Strain is defined as

$$\varepsilon = \frac{\Delta}{L} \tag{8.20}$$

where
    Δ is the elastic movement, which in this case is the anchor draw-in distance.
    L is the tendon length.

Young's modulus is

$$E = \frac{\sigma}{\varepsilon} \tag{8.21}$$

Combining these simple formulae provides the stress, or in this case the loss of prestress:

$$\sigma = \frac{\Delta E}{L} \tag{8.22}$$

Example 8.5: Anchorage draw-in

A manufacturer of an anchorage system specifies 5 mm draw-in for their anchorage system. Determine the loss of prestress if the tendons are 10 m long with Young's modulus of 210,000 N/mm².

From Equation 8.22, the loss of prestress is

$$\sigma = \frac{\Delta E}{L}$$

$$\sigma = \frac{5 \times 210000}{10000} = 105 \text{ N/mm}^2$$

8.2.4.2 Elastic shortening

The concrete will shorten under the compression force exerted by the tendons. If prestressed by only one tendon, then this shortening will not cause a drop in prestress; however, it is common for more than one tendon to be used, in which case the tendons jacked first will experience a loss of prestress due to the shortening induced by the tendons that are jacked subsequently.

The first tendon to be stressed will incur full elastic shortening losses, whereas the final tendon will incur no such losses. For simplicity, the average loss of prestress can be taken as 50% of the full loss due to elastic shortening, if multi-tendons are used.

In other words, the average loss of strain (ε_s) due to elastic shortening is equal to 50% of the concrete strain (ε_c), i.e:

$$\varepsilon_s = \frac{\varepsilon_c}{2} \tag{8.23}$$

Converting these strains to stress using Equation 8.21 provides the loss of stress

$$\sigma_{\text{loss}} = \frac{\sigma_c}{2} \times \frac{E_s}{E_c} \tag{8.24}$$

where
   σ_{loss} is the average loss of stress in the tendons.
   σ_c is the average stress in the concrete along the line of the tendons.
   E_c and E_s are Young's moduli for concrete and steel, respectively.

Example 8.6: Losses due to elastic shortening

A simply supported beam is 800 mm deep, 300 mm wide, spans 22 m and is prestressed with a force of 1400 kN. The eccentricity of the prestressing force is zero at the supports and 320 mm at midspan. Determine the loss of prestress due to elastic shortening if E_s = 210,000 N/mm^2 and E_c = 28,000 N/mm^2.

The UDL and corresponding midspan moment due to the self-weight are

$$w = 25 \times 0.8 \times 0.3 = 6 \text{ kN/m}$$

$$M = \frac{6 \times 22^2}{8} = 363 \text{ kN.m}$$

The area and second moment of area are

$$A = 800 \times 300 = 240000 \text{ mm}^2$$

$$I = \frac{300 \times 800^3}{12} = 1.28 \times 10^{10} \text{mm}^4$$

The tendons are located a distance e from the centroid. The stress in the concrete at a distance e from the neutral axis, due to the prestressing force P and applied moment M, is

$$\sigma = \frac{P}{A} + \frac{Pe^2}{I} - \frac{Me}{I} \tag{8.25}$$

At midspan, e = 320 mm; therefore,

$$\sigma = \frac{1400 \times 10^3}{240 \times 10^3} + \frac{1400 \times 10^3 \times 320^2}{1.28 \times 10^{10}} - \frac{363 \times 10^6 \times 320}{1.28 \times 10^{10}} = 7.96 \text{ N/mm}^2$$

At the supports, $e = 0$; therefore,

$$\sigma = \frac{1400 \times 10^3}{240 \times 10^3} + 0 - 0 = 5.83 \text{ N/mm}^2$$

The average stress is approximately equal to the average of the stress at the supports and midspan, i.e.,

$$\sigma = \frac{5.83 + 7.96}{2} = 6.89 \text{ N/mm}^2$$

And from Equation 8.24, the loss of prestress due to elastic shortening is approximately

$$\sigma_{loss} = \frac{\sigma_c}{2} \times \frac{E_s}{E_c} = \frac{6.89}{2} \times \frac{210000}{28000} = 25.8 \text{ N/mm}^2$$

A more exact solution could be gained by using a spreadsheet, although this illustrates the process.

8.2.4.3 Loss of prestress due to friction

Ducts are cast into members and steel tendons are passed through the ducts. When the tendons are stressed, a frictional force develops along their length and this reduces the tensile force exerted on the member. The tendons are usually draped; therefore, a frictional force develops along the inside of the drape due to the curvature (see Figure 8.10a). The ducts also have unintended imperfections in their profile, known as 'wobble', and this also induces additional friction (see Figure 8.10b). The resulting friction can be accounted for using the following loss factors:

$$\text{Curvature} = e^{-\mu\theta} \tag{8.26}$$

$$\text{Wobble} = e^{-\mu x K} \tag{8.27}$$

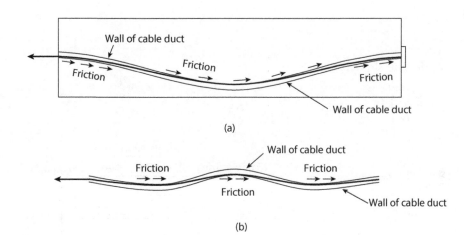

(a)

(b)

Figure 8.10 Types of friction loss. (a) Friction due to curvature and (b) section of a tendon illustrating friction due to 'wobble' of cable duct.

where

    μ is the coefficient of friction.

    θ is the total change in slope between the jacking point and the point where the tendon force is required.

    K is the wobble factor.

    x is the distance from the jack to where the tendon force is required.

Combining Equations 8.26 and 8.27

$$\text{Friction loss factor} = e^{-\mu(\theta + xK)} \tag{8.28}$$

The tendon force at distance x from the jack is determined by multiplying this by the jacking force.

Example 8.7: Calculation of friction losses in a continuous beam

Figure 8.11 shows a two-span beam that is 1000 mm wide and 340 mm deep. It is stressed with a prestressing force of 1000 kN; the cable ducts have a coefficient of friction $\mu = 0.2$ and a wobble factor $K = 0.01/\text{m}$. Determine the tendon force at Point A if the tendons are jacked from the far end, as illustrated in Figure 8.11.

The tendon will form a roughly parabolic shape, and the equation of a parabola of coordinates x, y and of span L and rise f is

$$y = \frac{4fx}{L} - \frac{4fx^2}{L^2}$$

Differentiating

$$\frac{dy}{dx} = \frac{4f}{L} - \frac{8fx}{L^2}$$

The slope is

$$\frac{dy}{dx} = \tan\theta$$

When $x = 0$, the end slope is

$$\theta = \tan^{-1}\left(\frac{4f}{L}\right) \tag{8.29}$$

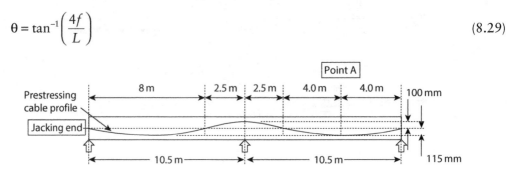

Figure 8.11 Continuous beam showing the dimensions of the prestressing tendon profile.

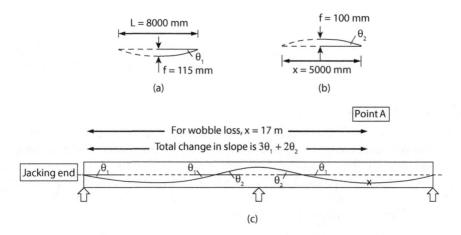

Figure 8.12 Determining end slope of prestressing strands. (a) Sagging regions, (b) hogging region over central support and (c) calculation of total change in slope between Point A and jacking end.

Considering the sagging region shown in Figure 8.12a

$$\theta_1 = \tan^{-1}\left(\frac{4 \times 115}{8000}\right) = 3.29^c$$

Repeating for the hogging region shown in Figure 8.12b

$$\theta = \tan^{-1}\left(\frac{4f}{L}\right)$$

$$\theta_2 = \tan^{-1}\left(\frac{4 \times 100}{5000}\right) = 4.57°$$

The total change in slope between the jack and Point A (see Figure 8.12c), is

$$\theta_{total} = 3\theta_1 + 2\theta_2 = 3 \times 3.29 + 2 \times 4.57 = 19.0°$$

or

$$\frac{19 \times \pi}{180} = 0.332 \text{ radians}$$

When calculating the loss due to wobble, x is the total length between the jack and Point A. From Equation 8.28

Loss factor = $e^{-0.2(0.332 + 17 \times 0.01)} = 0.904$

And the tendon force at Point A (excluding other losses) is

$$P = 1000 \times 0.904 = 904 \text{ kN}$$

8.2.4.4 Relaxation of the tendons

Relaxation in the steel will cause a small decline in the tendon force over a period of years. A total of 4% loss of tendon force due to relaxation would be typical during the lifetime of a structure, although tendon manufacturers and design standards specify precise figures.

8.2.4.5 Concrete creep

Elastic movements occur immediately concrete is stressed, in accordance with Hooke's law; however, concrete will continue to deform with time, with movement decreasing to zero over a period of years. This is called *creep* and prestressing induces stresses that will result in creep, which will in turn reduce the prestressing force.

Creep strain is a function of the stress in the concrete (σ_c), as well as the specific creep strain ($\varepsilon_{\text{specific}}$). The strain induced by creep is

$$\varepsilon = \sigma_c \varepsilon_{\text{specific}}$$

Since

$$E = \sigma/\varepsilon$$

The loss of stress in the tendons is

$$\sigma_{\text{loss}} = E_s \sigma_c \varepsilon_{\text{specific}} \tag{8.30}$$

where
σ_c is the concrete stress at the tendons.
$\varepsilon_{\text{specific}}$ is the specific creep strain, which can be calculated using codes of practice and should allow for the age of the concrete at stressing.

Example 8.8: Loss of prestress due to creep

Determine the loss of prestress for the beam described in Example 8.6 if the specific creep strain is 0.050×10^{-3} per N/mm^2.

From the previous working in Example 8.6, the average concrete stress was calculated approximately as 6.89 N/mm^2; therefore, from Equation 8.30 the loss of stress in the tendons is

$$\sigma_{\text{loss}} = E_s \sigma_c \varepsilon_{\text{specific}}$$

$$\sigma_{\text{loss}} = 210000 \times 6.89 \times 0.050 \times 10^{-3} = 72 \text{ N/mm}^2$$

8.2.4.6 Shrinkage

Concrete shrinks over a period of years after casting, due to loss of water from evaporation and the chemical bonding of water molecules during hydration of the cement. Shrinkage strain will reduce the stress in the tendons and the loss of stress is

$$\sigma_{\text{loss}} = \varepsilon_{\text{shrinkage}} E_s \tag{8.31}$$

The shrinkage strain ($\varepsilon_{shrinkage}$) can be determined from codes of practice and is dependent on factors such as the size of the member, the concrete grade, and type of aggregate.

Example 8.9: Calculation of prestress losses at transfer and SLS loading

A two-span continuous post-tensioned beam spans 20 m between simple supports and is 800 mm wide and 550 mm deep (see Figure 8.13). The tendons are jacked from both ends and the total prestress force is 4400 kN. Determine the short and long-term losses in prestress force at the central support.

Basic data. Tendon stress at transfer = 1200 N/mm^2, E_s = 210,000 N/mm^2, E_c = 28,000 N/mm^2, 4 mm anchorage draw-in, coefficient of friction of cable duct μ = 0.20 and wobble factor K = 0.01/m, shrinkage strain 100 $\times$ 10^{-6}, specific creep strain = 0.050 $\times$ 10^{-3} per N/mm^2, and there is a 2% relaxation of stress due to creep in the steel over time.

Anchorage draw-in. Friction between the strand and the cable duct will prevent the loss of prestress due to anchorage draw-in reaching the central support, so this loss is ignored.

Elastic shortening. The first step is to calculate the average stress due to bending and prestress along the line of the tendons. The UDL due to the self-weight is

$$w = 25 \times 0.8 \times 0.55 = 11 \text{ kN/m}$$

This beam can be considered as a propped cantilever. The first support moment is zero, because the beam is simply supported and the maximum sagging moment is

$$M = \frac{9wL^2}{128} = \frac{9 \times 11 \times 20^2}{128} = 309 \text{ kN.m}$$

and the internal support moment is

$$M = -\frac{wL^2}{8} = -\frac{11 \times 20^2}{8} = -550 \text{ kN.m}$$

The area and second moment of area of the beam are

$$A = 800 \times 550 = 440000 \text{ mm}^2$$

$$I = \frac{800 \times 550^3}{12} = 1.11 \times 10^{10} \text{ mm}^4$$

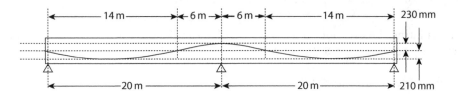

Figure 8.13 Continuous beam showing tendon profile.

From Equation 8.25, the stress at the first support (where e = 0) is

$$\sigma = \frac{4400 \times 10^3}{440 \times 10^3} + \frac{4400 \times 10^3 \times 0^2}{1.11 \times 10^{10}} - \frac{0 \times 0}{1.11 \times 10^{10}} = 10 \text{ N/mm}^2$$

And the maximum sagging stress (e = 210 mm) is

$$\sigma = \frac{4400 \times 10^3}{440 \times 10^3} + \frac{4400 \times 10^3 \times 210^2}{1.11 \times 10^{10}} - \frac{309 \times 10^6 \times 210}{1.11 \times 10^{10}} = 21.63 \text{ N/mm}^2$$

And the internal support stress (e = 230 mm) is

$$\sigma = \frac{4400 \times 10^3}{440 \times 10^3} + \frac{4400 \times 10^3 \times (-230)^2}{1.11 \times 10^{10}} - \frac{(-550) \times 10^6 \times (-230)}{1.11 \times 10^{10}} = 19.57 \text{ N/mm}^2$$

The average stress in the concrete along the profile of the tendon is approximately

$$\sigma = \frac{10 + 21.63 + 19.57}{3} = 17.1 \text{ N/mm}^2$$

And from Equation 8.24, the corresponding loss of prestress is

$$\sigma_{\text{loss}} = \frac{\sigma_c}{2} \times \frac{E_s}{E_c} = \frac{17.1}{2} \times \frac{210000}{28000} = 64.1 \text{ N/mm}^2$$

Friction losses. The question states that the tendons are stressed from both ends of the beam; therefore, only one of the two spans is considered during the friction loss calculations. From Equation 8.29, the end slope of the tendon at the outer supports is

$$\theta_1 = \tan^{-1}\left(\frac{4 \times 210}{14000}\right) = 3.43°$$

Repeating for the hogging region over the central supports

$$\theta_2 = \tan^{-1}\left(\frac{4 \times 230}{12000}\right) = 4.38°$$

The total change in slope between the jack and central support

$$\theta_{\text{total}} = 2\theta_1 + \theta_2 = 2 \times 3.43 + 4.38 = 11.24° = 0.196 \text{ rads}$$

When calculating the loss due to wobble, x is the total length between the jack and central support and from Equation 8.28

Friction loss factor = $e^{-\mu(\theta + xK)}$

Friction loss factor = $e^{-0.2(0.196 + 20 \times 0.01)} = 0.92$

And the loss of stress at the central support is

$$\sigma_{loss} = 1200 \times (1 - 0.92) = 96 \text{ N/mm}^2$$

Relaxation of the tendons. The loss of stress is

$$\sigma_{loss} = 1200 \times 0.02 = 24 \text{ N/mm}^2$$

Concrete creep. Assuming that the concrete stress remains unchanged over the life of the member, from Equation 8.30

$$\sigma_{loss} = 210000 \times 17.1 \times 0.050 \times 10^{-3} = 180 \text{ N/mm}^2$$

Shrinkage. From Equation 8.31

$$\sigma_{loss} = \varepsilon_{shrinkage} E_s = 100 \times 10^{-6} \times 210000 = 21 \text{ N/mm}^2$$

Losses at transfer. The combined losses from anchorage draw-in, elastic shortening, and friction are

$$\sum \sigma_{loss} = 60 + 96 = 156 \text{ N/mm}^2$$

Total long-term losses. The combined losses including losses at transfer in addition to creep in the steel, as well as concrete and shrinkage, are

$$\sum \sigma_{loss} = 156 + 24 + 180 + 21 = 381 \text{ N/mm}^2$$

8.2.5 Deflections

In conventional RC beams, it is impossible to accurately predict deflections because of the loss of stiffness due to cracking. However, Prestressed Concrete (PSC) beams can be designed to ensure no cracking occurs under SLS loads; therefore, accurate predictions are indeed possible.

Deflections need to be calculated at two separate stages in the life of a beam:

Stage 1. At 'transfer' (dead load only) (see Figure 8.14a)
Stage 2. Under full SLS loading (dead and imposed loads) (see Figure 8.14b)

Under Stage 1, the short-term value for Young's modulus and the α-loss factor are used, whereas under Stage 2, long-term Young's modulus and the β-loss factor apply. Early loading should quickly reduce the initial upwards deflections, although the creep deflection (due to the reduction of concrete's Young's modulus with time) will take 2 or more years to develop fully.

The calculation of deflection is split into two parts:

1. The uplift due to the tendon attempting to straighten, as illustrated in Figure 8.15b
2. Downwards deflection (sag) due to gravity loads; see Figure 8.15c

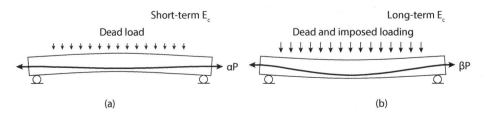

Figure 8.14 Short and long-term deflections. (a) Deflection 'at transfer' and (b) long-term (SLS) deflection.

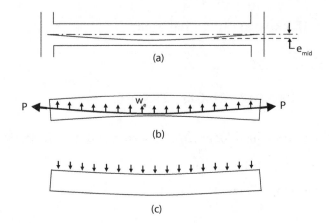

Figure 8.15 Simply supported beam. (a) Tendon profile, (b) uplift from tendon force and (c) sag from gravity loads.

8.2.5.1 Simply supported beam

The curved tendon shown in Figure 8.15a will try to straighten out. This will have a lifting effect, and the uplift force can be equated to a uniformly distributed load, w_e, shown in Figure 8.15b. The midspan moment is

$$M = \frac{w_e L^2}{8} \tag{8.32}$$

This moment is also equal to the tendon force multiplied by the eccentricity (Figure 8.15a), i.e.,

$$M = P e_{\text{mid}} \tag{8.33}$$

Combining Equations 8.32 and 8.33

$$w_e = \frac{-8 P e_{\text{mid}}}{L^2} \tag{8.34}$$

Provided the beam remains uncracked, the uplift deflection at midspan can now be calculated using the standard deflection equation for a beam with a UDL, i.e.,

$$\delta = \frac{5 w L^4}{384 E I} \tag{8.35}$$

Example 8.10: Deflection of a simply supported beam

A beam spans 20 m between simple supports and comprises a 0.8 m wide by 0.55 m deep PSC section subjected to an imposed serviceability load of 10 kN/m. The tendons are located on the centroid at the ends and 75 mm above the base of the beam (soffit) at midspan. Determine the following:

1. The deflection under the dead weight of the concrete at transfer (when tendons are first stressed)
2. The deflection under the dead and imposed loads (SLS loading)

Basic data
P = 6000 kN, α = 0.9, β = 0.8, Young's modulus of the concrete = 35,000 N/mm² at transfer and 14,000 N/mm² long term.

1. The second moment of area of the uncracked concrete section (ignoring rebar and tendons) is

$$I = \frac{800 \times 550^3}{12} = 11.1 \times 10^9 \, mm^4$$

From Equation 8.34, the uplift shown in Figure 8.15b, calculated using the short-term loss factor, is

$$w_e = \frac{-8\alpha Pe_{mid}}{L^2}$$

$$w_e = \frac{-8 \times 0.9 \times 6000 \times 0.200}{20^2} = -21.6 \, kN/m$$

From Equation 8.35 and using the short-term value of Young's modulus

$$\delta = \frac{5 \times (-)21.6 \times 20000^4}{384 \times 35000 \times 11.1 \times 10^9} = -116 \, mm$$

The beam self-weight is

$$w = 25 \times 0.8 \times 0.55 = 11 \, kN/m \, (= 11 \, N/mm)$$

From Equation 8.35, the corresponding sag is

$$\delta = \frac{5 \times 11 \times 20000^4}{384 \times 35000 \times 11.1 \times 10^9} = 59 \, mm$$

Finally, the total deflection at transfer is

$$\delta_{total} = -116 + 59 = -57 \, mm \uparrow$$

2. The uplift force now decreases as the long-term tendon losses are applied, from Equation 8.34

$$w_e = \frac{-8 \times 0.8 \times 6000 \times 0.200}{20^2} = -19.2 \text{ kN/m}$$

From Equation 8.35, the tendon force uplift (using the long-term value for Young's modulus) is

$$\delta = \frac{5 \times (-)19.2 \times 20000^4}{384 \times 14000 \times 11.1 \times 10^9} = -257 \text{ mm}$$

The SLS dead and imposed load is

$$w = 11 + 10 = 21 \text{ kN/m}$$

And the corresponding downwards deflection is

$$\delta = \frac{5 \times 21 \times 20000^4}{384 \times 14000 \times 11.1 \times 10^9} = 282 \text{ mm}$$

And finally, the total deflection under SLS loading is

$$\delta = -257 + 282 = +25 \text{ mm} \downarrow$$

8.2.5.2 Continuous beam

In this situation, the beams will have a tendon eccentricity at the supports (e_{end}) and at mid-span (e_{mid}) (see Figure 8.16a). The calculation of the uplift deflection from the tendon force is more complex than for the simply supported case. It is best calculated using the uplift from the total drape of the tendons (Figure 8.16b), minus the deflection due to end moments (Figure 8.16c).

δ_1 is the uplift due to the total drape of the tendon, $e_{mid} + e_{end}$ (see Figure 8.16b). This develops the following midspan moment

$$M = -P(e_{mid} + e_{end}) \tag{8.36}$$

Combining Equations 8.32 and 8.36

$$w_e = \frac{-8P(e_{mid} + e_{end})}{L^2} \tag{8.37}$$

And the corresponding deflection is calculated using Equation 8.35, which is for a simply supported beam, not a beam with fixed ends.

δ_2 is the downwards movement due to the support moments, shown in Figure 8.16c. The support moment is

$$M = Pe_{end} \tag{8.38}$$

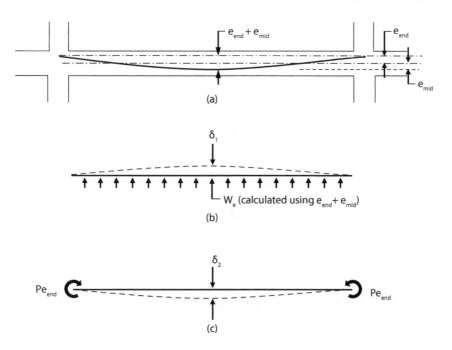

Figure 8.16 Tendon eccentric at supports. (a) Tendon profile, (b) deflection due to uplift from draped tendon and (c) deflection component due to support moment.

And the midspan deflection for a beam loaded with support moments, M, is

$$\delta = \frac{ML^2}{8EI} \qquad (8.39)$$

Combining Equations 8.38 and 8.39

$$\delta_2 = \frac{Pe_{end}L^2}{8EI} \qquad (8.40)$$

The final step is to calculate the downwards deflection due to gravity loads using the standard equation for a beam with *fixed supports*

$$\delta = \frac{wL^4}{384EI} \qquad (8.41)$$

In all of the above calculations, downwards deflection is taken as positive.

Example 8.11: Deflection in a continuous beam

Figure 8.17 shows a 0.24 m deep and 1.0 m wide continuous beam with fixed supports subjected to a uniformly distributed load. Determine the following:

1. The deflection under the dead weight of the concrete at transfer
2. The deflection under an SLS imposed load of 6 kN/m²

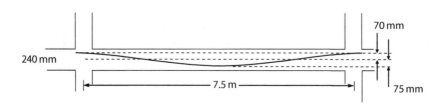

Figure 8.17 Continuous beams with approximately fixed supports.

Basic data
$P = 450$ kN, $\alpha = 0.9$, $\beta = 0.8$, $E_c = 35{,}000$ N/mm² at transfer and 14,000 N/mm² at the SLS

1. The second moment of area of the uncracked concrete section is

$$I = \frac{1000 \times 240^3}{12} = 1.152 \times 10^9 \, \text{mm}^4$$

From Equation 8.37, the uplift load from the tendon (with α-factor of 0.9 applied to tendon force) is

$$w_e = -\frac{8 \times 0.9 \times 450 \times 10^3 (75 + 70)}{7500^2} = -8.352 \, \text{N/mm}$$

From Equation 8.35 and using the short-term E

$$\delta_1 = \frac{5 \times (-8.352) \times 7500^4}{384 \times 35000 \times 1.152 \times 10^9} = -9 \, \text{mm}$$

And the end-moment deflection, from Equation 8.40, is

$$\delta_2 = \frac{0.9 \times 450 \times 0.07 \times 10^6 \times 7500^2}{8 \times 35000 \times 1.152 \times 10^9} = 5 \, \text{mm}$$

The self-weight is

$$w = 25 \times 0.24 \times 1.0 = 6 \, \text{kN/m}$$

Now, the self-weight deflection is calculated for a fixed end beam using Equation 8.41:

$$\delta = \frac{6 \times 7500^4}{384 \times 35000 \times 1.152 \times 10^9} = 1 \, \text{mm}$$

Finally, the combined deflection is

$$\delta = -9 + 5 + 1 = -3 \, \text{mm} \uparrow$$

i.e., the beam is expected to lift by 3 mm as soon as the tendons are stressed.

2. From Equation 8.37 and using the β factor (0.9) on the tendon force

$$w_e = \frac{-8 \times 0.8 \times 450 \times 10^3 (75 + 70)}{7500^2} = -7.424 \text{ N/mm}$$

From Equation 8.35 using long-term E

$$\delta_1 = \frac{5 \times (-7.424) \times 7500^4}{384 \times 14000 \times 1.152 \times 10^9} = -19 \text{ mm}$$

From Equation 8.40

$$\delta_2 = \frac{0.8 \times 450 \times 0.07 \times 10^6 \times 7500^2}{8 \times 14000 \times 1.152 \times 10^9} = 11 \text{ mm}$$

The UDL from the dead and imposed loadings is

$$w = 6 + 6 = 12 \text{ kN/m}$$

Equation 8.41

$$\delta = \frac{12 \times 7500^4}{384 \times 14000 \times 1.152 \times 10^9} = 6 \text{ mm}$$

Finally, the combined deflection under SLS loads is

$$\delta = -19 + 11 + 6 = -2 \text{ mm} \uparrow$$

8.2.5.3 Deflection in a propped cantilever

Figure 8.18 shows an edge beam (or slab) with one support simply supported and the other continuous. Deflections can be estimated using a similar approach to that described in Section 8.2.5.2, but with δ_1 determined using a reduced drape in the tendon ($e_{\text{mid}} + 0.5e_{\text{end}}$).

The midspan moment due to the tendon eccentricity is also

$$M = -P(e_{\text{mid}} + 0.5e_{\text{end}}) \tag{8.42}$$

Combing Equation 8.42 with Equation 8.32, the uplift load is

$$w_e = -\frac{8P(e_{\text{mid}} + 0.5e_{\text{end}})}{L^2} \tag{8.43}$$

The end moment from Figure 8.18 is

$$M = Pe_{\text{end}} \tag{8.44}$$

The midspan deflection in a beam with one end moment only is

$$\delta = \frac{ML^2}{16EI} \tag{8.45}$$

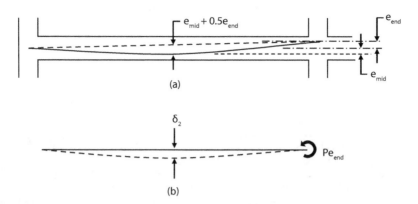

(a)

(b)

Figure 8.18 Unsymmetrical support conditions. (a) Tendon profile and (b) deflection due to downwards force from support moment.

Combining Equations 8.44 and 8.45

$$\delta_2 = \frac{Pe_{end}L^2}{16EI} \tag{8.46}$$

Finally, the dead load deflection is calculated using the standard propped cantilever with a UDL equation

$$\delta = \frac{wL^4}{185EI} \tag{8.47}$$

Example 8.12: Propped cantilever

A double-span continuous PSC beam spans 12 m between simple supports and is 450 mm wide and 500 mm deep (see Figure 8.19). The beam is subjected to a uniformly distributed imposed load of 12 kN/m. Determine the midspan deflection under

1. Self-weight at transfer
2. Serviceability limit state dead and imposed loads

Basic data
$P = 1100$ kN, $\alpha = 0.9$, $\beta = 0.8$, $E_c = 28,000$ N/mm² (short term) and 13,000 N/mm² (long term)

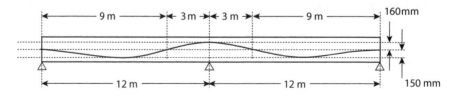

Figure 8.19 Two-span continuous beam.

1. The second moment of area of the section is

$$I = \frac{450 \times 500^3}{12} = 4.69 \times 10^9 \, \text{mm}^4$$

From Equation 8.43 and using the short-term loss factor ($\alpha = 0.9$)

$$w_e = -\frac{8 \times 0.9 \times 1100 \times 10^3 \times (150 + 160/2)}{12000^2} = -12.66 \, \text{N/mm}$$

From Equation 8.35 and using the short-term E

$$\delta_1 = \frac{5 \times (-12.66) \times 12000^4}{384 \times 28000 \times 4.69 \times 10^9} = -26 \, \text{mm}$$

From Equation 8.46 and using the short-term loss factor for tendon force

$$\delta_2 = \frac{0.9 \times 1100 \times 10^3 \times 160 \times 12000^2}{16 \times 28000 \times 4.69 \times 10^9} = 11 \, \text{mm}$$

The self-weight is

$$w = 25 \times 0.45 \times 0.5 = 5.6 \, \text{kN/m}$$

From Equation 8.47

$$\delta = \frac{5.6 \times 12000^4}{185 \times 28000 \times 4.69 \times 10^9} = 5 \, \text{mm}$$

And the total deflection at transfer is

$$\delta = -26 + 11 + 5 = -10 \, \text{mm} \uparrow$$

2. From Equation 8.43 and using the total loss factor ($\beta = 0.8$)

$$w_e = -\frac{8 \times 0.8 \times 1100 \times 10^3 \times (150 + 160/2)}{12000^2} = -11.25 \, \text{N/mm}$$

From Equation 8.35 and using the long-term E

$$\delta_1 = \frac{5 \times (-11.25) \times 12000^4}{384 \times 13000 \times 4.69 \times 10^9} = -50 \, \text{mm}$$

From Equation 8.46

$$\delta_2 = \frac{0.8 \times 1100 \times 10^3 \times 160 \times 12000^2}{16 \times 13000 \times 4.69 \times 10^9} = 21 \, \text{mm}$$

The UDL from the dead and imposed loads is

$$w = 5.6 + 12 = 17.6 \text{ kN/m}$$

From Equation 8.47

$$\delta = \frac{17.6 \times 12000^4}{185 \times 13000 \times 4.69 \times 10^9} = 32 \text{ mm}$$

Total deflection under SLS loads is

$$\delta = -50 + 21 + 32 = 3 \text{ mm} \downarrow$$

8.3 COMPOSITE CONSTRUCTION USING PRE-TENSIONED PRECAST CONCRETE BEAMS

Prestressed concrete beams are often used compositely with conventional RC. Examples are shown in Figures 8.1 and 8.20, in which pre-tensioned beams support concrete slabs. The basic principles of ULS design are the same as for ordinary prestressed concrete; however, the checks for the SLS condition are slightly different. As with the conventional steel–concrete composite construction, the beams support the wet weight of the concrete if unpropped, with all loads thereafter supported by the combined action of the precast concrete section and the conventional concrete.

A freshly cast slab, such as that shown in Figure 8.20, will initially have a lower Young's modulus than the older precast sections due to the difference in age. In addition, Young's modulus is dependent on the concrete grade. The precast manufacturer may choose a high grade concrete because of the early strength gain it will provide. Thus the slab and the precast units may have significantly different values of Young's modulus. The effect on stress of different Young's modulus values can be assessed using the *method of transformed sections*. To do this the *modular ratio* is required, which is the ratio of Young's moduli For the two different grades of concrete.

If the composite action second moment of area is calculated using uncracked properties for the concrete, the SLS checks must include a check that the tensile stresses are within the permissible tensile stress limits. The shear strength of the contact face between the concrete elements also needs assessing, with steel shear links used to transfer the shear forces required to develop the composite action.

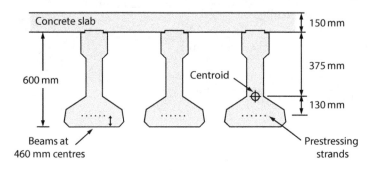

Figure 8.20 Cross-section through a PSC bridge deck at midspan.

Example 8.13: Stress calculations for a composite bridge deck

Figure 8.20 shows a bridge that spans 13.5 m between simple supports. The pre-tensioned beams support the concrete slab during casting, i.e., they are unpropped during casting. Once hardened, both the slab and the beams support loads through composite action. Determine the stress distribution in the cross section at midspan at the following stages:

1. At transfer, but before the concrete slab has been cast
2. When supporting the wet weight of the slab
3. When supporting an imposed load of 9 kN/m per beam

Basic data

Each beam has self-weight of 3.26 kN/m, cross-sectional area of 0.136m², second moment of area of 6.5×10^9 mm⁴ and is pretensioned with a force (after losses) of 1000 kN. The pre-tensioning strands are located 130 mm below the centroid, as shown in Figure 8.20. The (modular) ratio between the Young's moduli of the slab over beam is 0.667.

1. The midspan moment due to the 3.26 kN/m self-weight is

$$M = \frac{3.26 \times 13.5^2}{8} = 74.3 \text{ kN.m}$$

The section moduli for the top and bottom of the precast concrete sections are

$$Z_T = \frac{6.5 \times 10^9}{375} = 17.3 \times 10^6 \text{ mm}^3$$

$$Z_B = \frac{6.5 \times 10^9}{225} = 28.8 \times 10^6 \text{ mm}^3$$

The tendon eccentricity (i.e. the distance between the tendons and centroid) is 130 mm, and from Equation 8.1 the stress at the top of the section at the interface with the slab soffit is

$$\sigma_T = \frac{P}{A} - \frac{Pe}{Z_T} + \frac{M}{Z_T}$$

$$\sigma_T = \frac{1000 \times 10^3}{0.136 \times 10^6} - \frac{1000 \times 10^3 \times 130}{17.3 \times 10^6} + \frac{74.3 \times 10^6}{17.3 \times 10^6}$$

$$\sigma_T = 7.4 - 7.5 + 4.3 = +4.2 \text{ N/mm}^2$$

And from Equation 8.2 at the bottom

$$\sigma_B = \frac{P}{A} + \frac{Pe}{Z_B} - \frac{M}{Z_B}$$

$$\sigma_B = 7.4 + 4.5 - 2.6 = +9.3 \text{ N/mm}^2$$

2. The dead weight of the concrete slab is

$$w = 25 \times 0.46 \times 0.15 = 1.72 \text{ kN/m}$$

And the corresponding midspan moment per beam is

$$M = \frac{1.72 \times 13.5^2}{8} = 39.2 \text{ kN.m}$$

The additional bending stresses are now added to those from Step 1, i.e.,

$$\sigma_T = +4.2 + \frac{M}{Z_T} = 4.2 + \frac{39.2 \times 10^6}{17.3 \times 10^6} = +6.5 \text{ N/mm}^2$$

$$\sigma_B = +9.3 - \frac{M}{Z_B} = +7.9 \text{ N/mm}^2$$

3. The midspan moment due to the imposed load is

$$M = \frac{9 \times 13.5^2}{8} = 205 \text{ kN.m}$$

This is resisted by the combined action of the precast concrete sections and the slab. The top slab has a lower value of Young's modulus than the precast concrete beams; therefore the *method of transformed sections* must be used to determine the *effective width* of the top flange. The effective width of the 460 mm wide slab is determined using the *modular ratio* (0.667), i.e.,

Eff. width = $460 \times 0.667 = 306.7$ mm

Taking moments of area about the top of the slab to determine the position of the centroid of area

$$\bar{y} = \frac{306.7 \times 150^2 \times 0.5 + 0.136 \times 10^6 \times (375 + 150)}{306.7 \times 150 + 0.136 \times 10^6} = 411 \text{ mm}$$

Using the *parallel axis theorem*, the second moment of area is

$$I = \frac{306.7 \times 150^3}{12} + 306.7 \times 150 \times (411 - 75)^2 + 6.5 \times 10^9 + 0.136 \times 10^6 (375 + 150 - 411)^2$$

$$I = 13.5 \times 10^9 \text{ mm}^4$$

At the top of the slab

$$\sigma = \text{mod. ratio} \times \frac{M_y}{I}$$

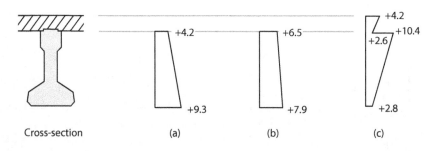

Figure 8.21 Stress distribution at the different stages, in N/mm². (a) At transfer, (b) with wet concrete and (c) SLS loading.

$$\sigma = 0.667 \times \frac{205 \times 10^6 \times 411}{13.5 \times 10^9} = +4.2 \text{ N/mm}^2$$

And at the bottom of the slab

$$\sigma = 0.667 \times \frac{205 \times 10^6 \times (411 - 150)}{13.5 \times 10^9} = +2.6 \text{ N/mm}^2$$

At the top of the precast concrete beams the imposed load stresses are added to the previously calculated values

$$\sigma = +6.5 + \frac{205 \times 10^6 (411 - 150)}{13.5 \times 10^9} = +10.4 \text{ N/mm}^2$$

And at the bottom of the beams

$$\sigma = +7.9 - \frac{205 \times 10^6 \times (600 + 150 - 411)}{13.5 \times 10^9} = +2.8 \text{ N/mm}^2$$

The build-up of the stresses at transfer, supporting the wet weight of the concrete and under the imposed loads is sketched in Figure 8.21.

8.4 ULS BENDING STRENGTH

The basic method for determining the ultimate moment capacity of a PSC beam is the same as that used for the design of conventionally reinforced beams, as described in Chapter 7. However, there are subtle differences related to the greater strength of prestressing tendons in comparison with conventional rebar. The basic design assumptions regarding the assumed stress distribution at the ULS are sketched in Figure 8.22b.

The compressive force in the concrete is

$$C_c = f_{cd} \times b \times 0.8x$$

where
   x is the neutral axis depth
   b is the width of the beam (see Figure 8.22a).

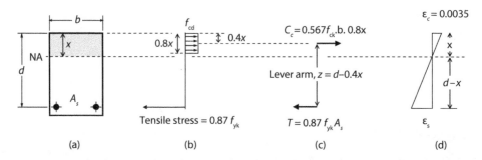

Figure 8.22 Moment capacity assumptions. (a) Cross-section, (b) design stress distribution, (c) design forces and (d) strain distribution.

Inputting the design strength of concrete from Equation 7.1, this becomes

$$C_c = 0.567 f_{ck} \times b \times 0.8x \tag{8.48}$$

From Equation 7.4, the tensile strength of the steel is

$$T = 0.87 f_{yk} A_s$$

If no external axial force is applied, then horizontal equilibrium tells us that

$$T = C_c$$

The lever arm distance between the steel and the concrete (Figure 8.22c) is

$$z = d - 0.4x \tag{8.49}$$

And the moment capacity is

$$M = Tz \tag{8.50}$$

As with any RC beam (Equation 7.8), the neutral axis depth is limited to not greater than $0.45d$, in order to ensure that the beam has sufficient ductility.

In PSC beams, it is also necessary to check that the steel tendons have yielded. This is because the yield stress of prestressing steel is very high, so there is a real risk that the concrete will crush before the tendons have yielded. The maximum compression strain (ε_c) acceptable in normal strength concrete is 0.0035. From similar triangles (Figure 8.22d), the bending strain in the steel (ε_s) is

$$\frac{\varepsilon_s}{d - x} = \frac{0.0035}{x} \tag{8.51}$$

The pre-tensioning strain is

$$\varepsilon_{pre} = \frac{\beta \sigma_{pre}}{E} \tag{8.52}$$

where σ_{pre} is the prestress. The total strain in the tendons at the ULS is the bending strain (σ_s) + pre-tensioning strain (σ_{pre}). If the total strain $\geq$ yield strain, then yielding occurs and the design assumptions hold true, where the yield strain is

$$\varepsilon_y = \frac{f_{yk}}{E}$$

If total strain < yield strain, then the stress in the tendons will remain below the yield stress when the concrete fails in crushing. In that case, the moment capacity must be recalculated accounting for the reduced stress in the steel.

Example 8.14: Moment capacity calculation

A 250 mm wide, 600 mm deep beam is reinforced with 2 × 25 mm diameter conventional rebar, in addition to 4 × 10 mm diameter pre-tensioning tendons. The centroids of the reinforcements are all located 50 mm above the soffit of the beam. The tendons are initially stressed at 1200 N/mm², although the yield stress is 1600 N/mm². Determine the ULS moment capacity.

Basic data
Conventional rebar A_s = 982 mm², f_{yk} = 500 N/mm²; f_{ck} = 35 N/mm², crushing strain = 0.0035. Prestressing tendons A_s = 314 mm², β = 0.8; E_s = 210,000 N/mm².

Effective depth, d = 600 – 50 = 550 mm and rearranging Equation 8.48 in terms of x

$$x = \frac{C_c}{0.567 \times 35 \times 250 \times 0.8}$$

$$x = \frac{C_c}{3969}$$

The combined tensile strength of the rebar and prestressing tendons from Equation 7.4 is

$$T = (1600 \times 314 + 500 \times 982) \times 0.87 \times 10^{-3} = 864 \text{ kN}$$

From horizontal equilibrium, $C_c = T$; therefore

$$x = \frac{864 \times 10^3}{3969} = 218 \text{ mm}$$

This is less than the limit of x = 0.45d (248 mm); therefore the section should be ductile. From Equation 8.49, the lever arm is

$$z = d - 0.4x$$

$$z = 550 - 0.4 \times 218 = 462.8 \text{ mm}$$

And from Equation 8.50, the moment capacity is

$$M = Tz = 864 \times 10^3 \times 462.8 \times 10^{-6} = 400 \text{ kN.m}$$

This assumes that the prestressing steel has yielded, although this needs to be checked. From Equation 8.51, the bending strain is

$$\frac{\varepsilon_s}{d-x} = \frac{0.0035}{x}$$

$$\varepsilon_s = \frac{0.0035(550-218)}{218} = 0.0053$$

From Equation 8.52, the pre-strain is

$$\varepsilon_{pre} = \frac{\beta\sigma_{pre}}{E}$$

$$\varepsilon_{pre} = \frac{1200\times0.8}{210000} = 0.0046$$

Total strain = 0.0047 + 0.0053 = 0.0100

The yield strain is

$$\text{Yield strain} = \frac{f_{yk}}{E} = \frac{1600}{210000} = 0.0076$$

Since the total strain (0.0100) > yield strain (0.0076), the tendons will yield and the moment calculation is correct.

8.5 ULS SHEAR STRENGTH

The basic theory for the design of shear reinforcement for prestressed concrete is the same as that for conventional RC, as described in Chapter 7. However, axial compression from prestressing can enhance the shear strength. If uncracked because of prestressing, then shear strength will be enhanced further. This section considers the theory behind the estimation of the shear strength for uncracked beams *without* conventional shear reinforcement.

Figure 8.23a shows an element of concrete subjected to a compression stress due to pre-stress (σ_c) and a shear stress (τ). Mohr's circle of stress (Figure 8.23b) tells us that the centre of the circle is located at a distance $\sigma_c/2$ from the y axis. The radius of Mohr's circle (R) is

$$R = \frac{\sigma_c}{2} - \sigma_t$$

From Pythagoras

$$R^2 = \left(\frac{\sigma_c}{2}\right)^2 + \tau^2$$

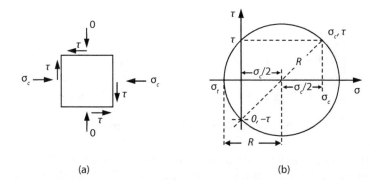

(a) (b)

Figure 8.23 Mohr's circle of stress. (a) Element under compression and shear and (b) Mohr's circle of stress.

Combining these equations

$$\left(\frac{\sigma_c}{2} - \sigma_t\right)^2 = \left(\frac{\sigma_c}{2}\right)^2 + \tau^2$$

which simplifies to

$$\tau = \sqrt{\sigma_t^2 - \sigma_c\sigma_t} \tag{8.53}$$

where
σ_t is the principal tensile stress (negative for tension).
τ is the shear stress.

The shear stress (τ) a distance y from the neutral axis (see Figure 8.24) is

$$\tau = \frac{VA'\overline{y}}{b_o I} \tag{8.54}$$

where I is the second moment of area of the cross section and V is the applied shear force. Combining Equations 8.53 and 8.54 provides the shear strength of the (uncracked) PSC beam without shear reinforcement, i.e.,

$$V = \frac{b_o I}{A'\overline{y}} \sqrt{\sigma_t^2 - \sigma_c\sigma_t} \tag{8.55}$$

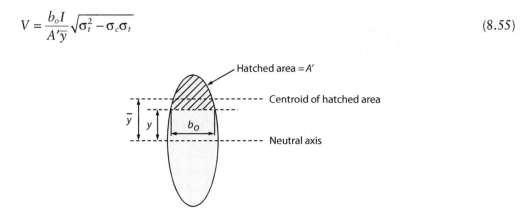

Figure 8.24 Notation used in the shear equation.

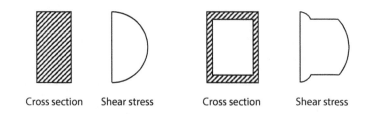

Cross section   Shear stress   Cross section   Shear stress

Figure 8.25 Distribution of shear stress.

This calculation assumes that the shear capacity limit is reached when the concrete first cracks, i.e., when σ_t reaches the tensile strength. Figure 8.25 shows the distribution of shear stress for beams of rectangular and box-shaped cross sections. These show that the maximum shear stress occurs at the centroid. For this reason, the critical shear strength calculation is performed at the centroid.

Example 8.15: Shear strength of a prestressed concrete unit

Figure 8.26 shows a Prestressed Concrete unit. The total combined force from all the pretensioning cables $P = 18$MN, and the combined eccentricity from all the tendons is 233 mm below the centroid. If the units are uncracked and the maximum tensile stress in the concrete is limited to 2.5 N/mm², determine the maximum shear force that the units can resist in the absence of shear reinforcement.

1. The second moment of area and cross-sectional area are

$$I = \frac{2100 \times 2800^3}{12} - \frac{1800 \times 2000^3}{12} = 2.64 \times 10^{12} \text{ mm}^4$$

$$A = 2100 \times 2800 - 2000 \times 1800 = 2.28 \times 10^6 \text{ mm}^2$$

The stress due to the 18MN prestress at the centroid is

$$\sigma = \frac{P}{A} + \frac{Pey}{I}$$

$$\sigma = \frac{18 \times 10^6}{2.28 \times 10^6} + \frac{18 \times 10^6 \times 233 \times 0}{2.64 \times 10^{12}} = 7.89 \text{ N/mm}^2$$

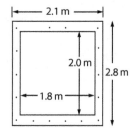

Figure 8.26 Cross section through a PSC rectangular tube.

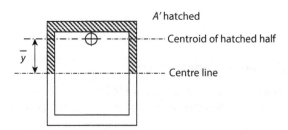

Figure 8.27 Design parameters for calculation of shear stress.

And taking moments of area about the centre line as shown in Figure 8.27

$$300 \times 1400 \times 700 + (2100 - 300) \times 400 \times (1400 - 200) = \bar{y} \times 0.5 \times 2.28 \times 10^6$$

$$\bar{y} = 1016 \text{ mm}$$

And from Equation 8.55, the shear strength of the uncracked section is

$$V = \frac{b_o I}{A' \bar{y}} \sqrt{\sigma_t^2 - \sigma_c \sigma_t}$$

$$V = \frac{(2100 - 1800) \times 2.64 \times 10^{12}}{0.5 \times 2.28 \times 10^6 \times 1016} \times 10^{-3} \sqrt{(-2.5)^2 - 7.89 \times (-2.5)} = 3485 \text{ kN}$$

8.6 DESIGN OF ANCHORAGES

The anchorage imposes large stress concentrations to the ends of members and this can lead to splitting. This can be prevented by the use of anchorage reinforcement, which can be designed using the *strut and tie* method described in Chapter 9. During design, the stress in the rebar should not be greater than 60% of the yield stress. This keeps the rebar strain low and helps to control crack widths.

The exact form of the strut and tie model allows for some variation, although a dispersion of 1:2 is commonly used for simplicity. An example of the simplest strut and tie model is shown in Figure 8.28. This section provides only a brief introduction into the principles of designing anchorage, without reference to the checking of stresses in the concrete. These can

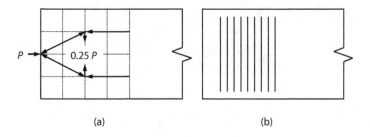

(a)   (b)

Figure 8.28 Examples of strut and tie models for anchorages. (a) Strut and tie model and (b) reinforcement detail.

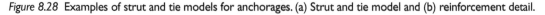

be checked using the full strut and tie method presented in Chapter 9, and more variations in strut and tie models are presented in Nawy (2009).

Example 8.16: Single tendon anchorage

A 400 mm deep beam is post-tensioned with a force of 500 kN using one tendon located on the centroid. Design the anchorage reinforcement if $f_{yk} = 500$ N/mm².

Figure 8.28a shows the strut and tie model developed for this question. A 2:1 load dispersion is assumed; therefore, from equilibrium the splitting force is

$$T = \frac{500}{4} = 125 \text{ kN}$$

The stress in the steel is limited to 60% of the yield stress; therefore, by adjusting Equation 7.4

$$T = 0.6 \times 0.87 \ A_s f_{yk} \tag{8.56}$$

The steel area is

$$A_s = \frac{125 \times 10^3}{0.6 \times 0.87 \times 500} = 479 \text{ mm}^2$$

The rebar are provided in a hoop shape around the tendon anchor; it follows that each bar passes through the crack twice. Therefore, 9 × 6 mm diameter links provide 2 × 254 mm² > 479 mm², which is thus adequate (see Table 8.1 and Figure 8.28b).

Example 8.17: Two tendon anchorage

A 400 mm deep beam is post-tensioned using two tendons, each of which is located 120 mm to either side of the centroid. Each tendon is prestressed with a tendon capable of loading the beam with a force of 260 kN at the ULS. Design the anchorage reinforcement if $f_{yk} = 500$ N/mm².

Figure 8.29a shows the two types of cracking that can occur, which include splitting between the two point loads and bursting of the concrete behind the bearing plates. Figure 8.29b shows the strut and tie model for splitting. From equilibrium and assuming a 1:2 slope of the diagonal struts, the tension force is

$$T = \frac{260}{\tan(63.4)} = 130 \text{ kN}$$

Table 8.1 Cross-sectional areas of groups of rebar, in mm²

Diameter	Rebar Number								
	1	2	3	4	5	6	7	8	9
6	28	57	85	113	141	170	198	226	254
12	113	226	339	452	556	679	792	905	1020

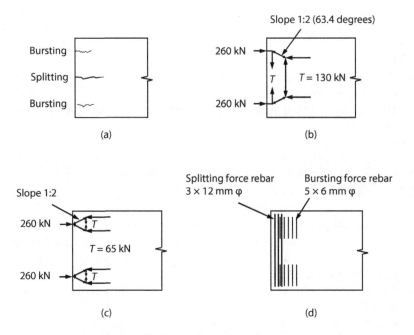

Figure 8.29 Beam resisting two point loads from prestressing. (a) Crack patterns, (b) splitting force strut and tie model, (c) bursting force strut and tie model and (d) rebar arrangement.

From Equation 8.56, the area of steel needed to prevent splitting is

$$A_s = \frac{130 \times 10^3}{0.6 \times 0.87 \times 500} = 498 \text{ mm}^2$$

Provide 3 × 12 mm diameter links – this gives a total cross-sectional area of 2 × 339 = 678 mm² (see Table 8.1). The anti-bursting strut and tie model is shown in Figure 8.29c. From equilibrium, T = 65 kN; therefore,

$$A_s = \frac{65 \times 10^3}{0.6 \times 0.87 \times 500} = 249 \text{ mm}^2$$

Provide 5 × 6 mm diameter rebar = 2 × 141 = 282 mm² (see Table 8.1). The final rebar arrangement is sketched in Figure 8.29d.

Problems

Solutions to these problems are provided at https://www.crcpress.com/9781498741217

P.8.1. A beam is 300 mm deep and 200 mm wide and spans 4.6 m between simple supports. It supports a UDL of 9.7 kN/m. If a prestress force of 200 kN is applied to the section, with a maximum eccentricity of the tendon of 70 mm below the centroid, determine the stress distribution at the top and bottom of the beam at midspan under unfactored loads.
   Ans. +8.5 N/mm² top and –1.9 N/mm² bottom.

P.8.2. A post-tensioned RC beam is 0.5 m wide and spans 14 m between simple supports. It supports a UDL of 14 kN/m and the tendon is located 0.15 m below the
centroid at midspan. The limiting stresses and anchorage loss factors are listed
below.
   a. Determine the minimum beam depth to resist the applied SLS loads without
      cracking or crushing.
   b. If the beam depth is set at 0.6 m, determine the maximum and minimum limits
      on tendon force.
   Basic data
   $\alpha = 0.9$, $\beta = 0.8$, $\sigma_{min,t} = 0$, $\sigma_{min,sls} = 0$, $\sigma_{max,t} = 20$ N/mm², $\sigma_{max,sls} = 16$ N/mm²
   Ans. (a) $h = 0.52$ m and 0.495 m and (b) $P \le 4083$ kN, $P \le 3483$ kN, $P \ge 1169$ kN,
   $P \ge 2634$ kN, $\rightarrow 2634$ kN $\le P \le 3483$ kN.

P.8.3. A 500 mm deep and 700 mm wide beam spans 18 m between simple supports and
is prestressed with tendons that develop a combined force of 3500 kN. The eccentricity of the tendons at the end of the beams is zero and at midspan is 170 mm.
$E_c = 28,000$ N/mm², $E_s = 210,000$ N/mm².
   a. Determine the stress in the concrete at the tendon position at the supports and
      at midspan.
   b. Determine the loss of prestress due to elastic shortening.
   Ans. (a) 10.0 N/mm² and 15.6 N/mm² and (b) 48.0 N/mm².

P.8.4. A 220 mm deep and 1000 mm wide beam is shown in Figure 8.30.

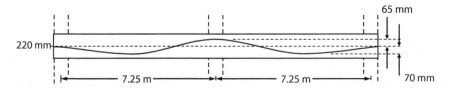

Figure 8.30 Two-span continuous beam.

   a. Determine the midspan deflection at transfer (dead load only).
   b. Determine the midspan deflection under an SLS imposed load of 8 kN/m.
   Basic data
   $P = 550$ kN, $\alpha = 0.9$, $\beta = 0.8$, $E_c = 35,000$ N/mm² short term, $E_c = 14,000$ N/mm²
   long term
   Ans. (a) –2.86 mm and (b) 3.9 mm.

P.8.5. A bridge deck comprises inverted T-beams that span 16 m between simply supported bearings (see Figure 8.31). Using the basic data provided below determine
the following:
   a. The stress distribution in the inverted T-beam at transfer.
   b. The stress distribution when resisting the wet weight of the concrete.
   c. The stress distribution supporting an imposed load of 8 kN/m.
   Basic data for the inverted T-beams
   Prestress force = 1200 kN, I = 8.6×10^9 mm⁴, cross-sectional area = 0.16 m²,
   concrete density = 25 kN/m³

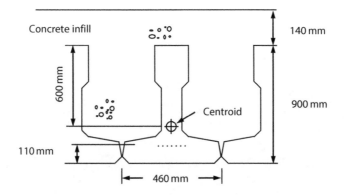

Figure 8.31 Prestressed composite bridge deck.

Ans. (a) 0.55 and 10.98 N/mm² top and bottom, (b) 18.4 and 2.1 N/mm² and (c) +3.1 N/mm² top of bridge deck, 20.6 and –1.0 N/mm² top and bottom of inverted T-section.

P.8.6. A 160 mm wide and 220 mm deep beam is post-tensioned with 2 × 8 mm diameter tendons located 30 mm above the soffit. The tendons are initially stressed at 1200 N/mm², although the yield stress is 1600 N/mm².
1.   Determine the ULS moment capacity.
2.   Determine the strain in the tendons at the ULS.
Basic data
f_{ck} = 40 N/mm², E_s = 210,000 N/mm², concrete crushing strain = 0.0035, β = 0.8, A_s = 100.5 mm²
Ans. (a) 23.9 kN.m and (b) 0.015.

P.8.7. Figure 8.32 shows prestressed concrete units. The total combined force from all the prestress tendons is 13 MN and the combined eccentricity of the tendons is 207 mm below the centroid. If the units are uncracked and the maximum tensile stress in the concrete is limited to 4 N/mm², determine the maximum applied shear force that the units can resist.

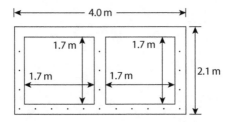

Figure 8.32 Prestressed concrete units.

Ans. 6212 kN.

REFERENCE

Nawy, E., 2009. *Prestressed Concrete: A Fundamental Approach.* 5th Edition. USA: Pearson.

Chapter 9

Strut and tie modelling of reinforced concrete

This chapter describes strut and tie modelling (STM) of RC. Finite Element Analysis (FEA) analysis cannot easily deal with cracking that normally occurs in RC structures. In contrast, STM provides safe estimates of strength irrespective of cracking. The problems of FEA modelling of concrete were illustrated by the Sleipner-A oil rig, which sank whilst being lowered into a Norwegian fiord. When it hit the seabed, it caused an earthquake recorded at 3.0 on the Richter scale and cost $700 million. The investigators found that the failure could have been avoided if STM had been used in conjunction with FEA.

9.1 INTRODUCTION TO STM

Elastic beam theory becomes inaccurate when the span-to-depth ratio of a beam is less than 4. This problem is illustrated by the deep beam sketched in Figure 9.1. Elastic beam theory would forecast a linear distribution of bending stresses as shown in Figure 9.1a, whereas FEA would predict the stresses shown in Figure 9.1b. In practice, tensile stresses would cause the concrete to crack. The tensile forces previously carried by the concrete would transfer to the steel reinforcement located along the base of the beam. The arrangement of struts and ties shown in Figure 9.1c provides a realistic representation of the cracked behaviour, where compression struts are indicated by the letter C and the tension force in the rebar is indicated by the letter T.

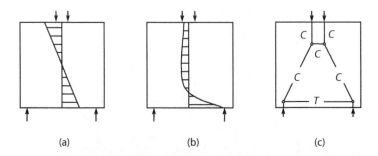

Figure 9.1 Midspan bending stress distributions. (a) Bernoulli's stress distribution, (b) stress distribution from elastic FEA and (c) notional arrangement of struts and ties.

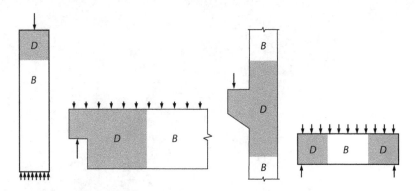

Figure 9.2 Illustration of B-regions and D-regions.

The STM method is used for modelling areas of stress concentration within structures. These are termed 'D-regions', with *D* for *discontinuity*. Areas where conventional beam theory proves accurate are termed 'B-regions', where *B* stands for *beam*. The D-regions are generally assumed to be square in proportion, i.e., equal in length to the member depth (see Figure 9.2).

The D-regions result from either geometric discontinuities (Figure 9.3) or from concentrated loading (Figure 9.4). Typical examples of D-regions include connections, corners, openings in beams, and deep beams, such as pile caps. The STM is used to design the reinforcement within the D-regions, whereas standard theory is used for the B-regions.

9.2 FORMULATION OF THE STRUT AND TIE MODEL

The positioning of the struts and ties should ideally correspond with the natural flow of stresses in cracked RC. Figure 9.5a shows the flow of compression stress away from a point load. Below the D-region, the concentrated stresses under the load become uniformly distributed. This uniform distribution can be represented by two compression struts positioned centrally in the two halves of the member, labelled C1 in Figure 9.5b. The angle of the diagonal compression struts labelled C2 can vary, although an easy and safe approximation is to assume a 1:2 slope. This angle makes the calculation of the tie force particularly easy, with resolution of the forces showing that the tie force $T1 = P/4$ (see Figure 9.5b).

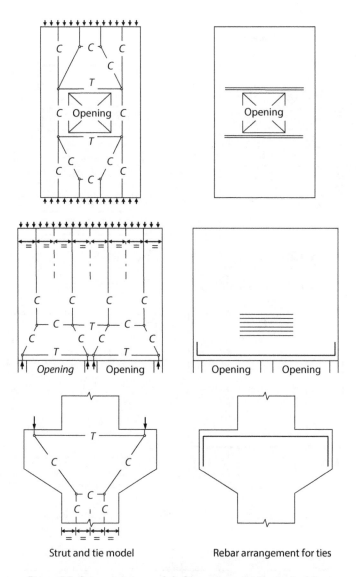

Strut and tie model   Rebar arrangement for ties

Figure 9.3 Strut and tie models for geometric discontinuities.

Rebars are positioned on the centreline of the ties. If more than one rebar is used, then the centre of the group of bars should be located on the centreline of the tie, as is the case for the rebar shown in Figure 9.5c, which is spread to either side of the centre line of $T1$.

The standard models shown in Figure 9.6 can be adapted to cover a wide range of practical situations. When deciding upon which model to use, the designer should select one that involves the lowest strain energy, where

$$\text{strain energy} = \text{force} \times \text{length} \times \text{strain}$$

Strain energy is highest in ties, because rebar strain tends to be much higher than the strain in the concrete struts. Therefore, the *length of the ties should be minimised* when deciding which arrangement of struts and ties to use.

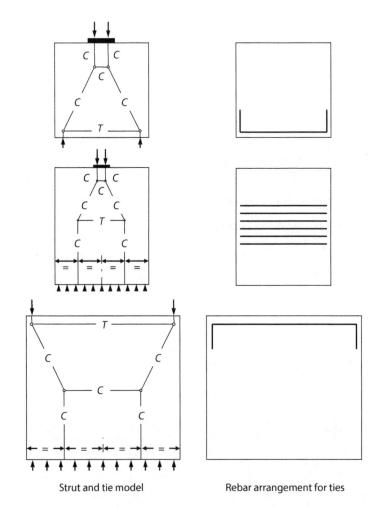

Strut and tie model   Rebar arrangement for ties

Figure 9.4 Strut and tie models for concentrated loads.

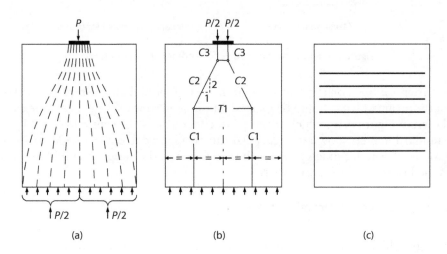

(a)   (b)   (c)

Figure 9.5 Formulation of STM for an element with concentrated load. (a) Flow of compression stress, (b) the strut and tie model and (c) rebar distribution for *T*1.

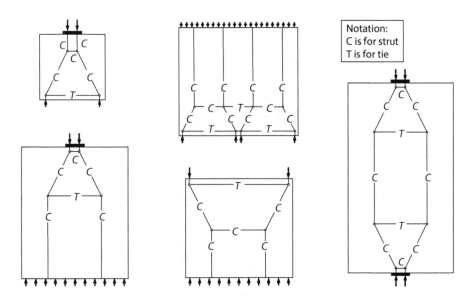

Figure 9.6 Common strut and tie models.

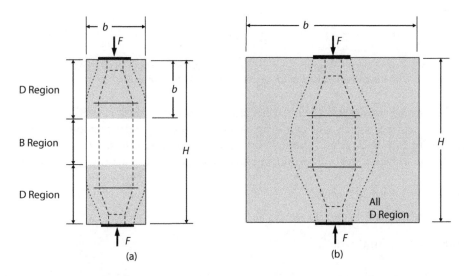

Figure 9.7 Design parameters for struts with partial and full discontinuities. (a) Strut with partial discontinuity, $b < H/2$ and (b) strut with full discontinuity, $b > H/2$.

Figure 9.7a shows a member composed of B- and D-regions because $b < H/2$. This is called a *partial discontinuity*, whereas if $b > H/2$ (Figure 9.7b) the entire member is a D-region and this is called a *full discontinuity* member. The classification of full or partial discontinuity has design implications when defining the forces in the struts and ties.

9.2.1 Partial discontinuity

The positions and forces in the members can be defined by considering the diagrams in Figure 9.8, which can represent the D-regions from Figure 9.7a. The transverse stress

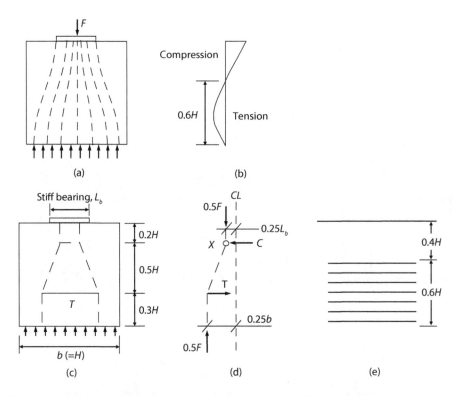

Figure 9.8 A partial discontinuity member. (a) Stress trajectories, (b) transverse stresses, (c) strut and tie model, (d) idealised forces one side and (e) position of rebar.

distribution is shown in Figure 9.8b. This shows tension stresses caused by the spreading compression stresses, which will cause cracking. The node positions shown in Figure 9.8c are taken from Eurocode 2 and the forces in the members are shown in the free body diagram sketched in Figure 9.8d. Considering moment equilibrium around point X

$$0.5F \times \left(\frac{b}{4} - \frac{L_b}{4} \right) = T \times 0.5b$$

Therefore, the force in the tie is

$$T = \frac{F}{4} \left(\frac{b - L_b}{b} \right) \tag{9.1}$$

Figure 9.8e shows the positioning of the rebar. These are represented by the force T in the STM. According to the theory, the zone directly below the point load (0.4H) can go unreinforced; however it would be inadvisable to not reinforce this area, because laboratory tests show that the crack will grow and extend higher. Therefore, the reinforcement should be extended above to control cracking.

9.2.2 Full discontinuity

Figure 9.9a shows a member that is entirely comprises a D-region in which a concentrated load creates tensile stresses in the central region (Figure 9.9b). The geometry of the strut and

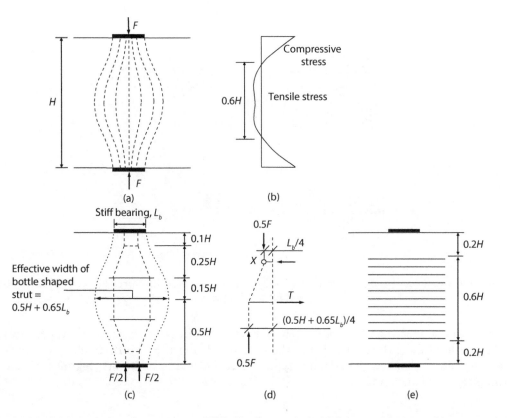

Figure 9.9 A full discontinuity member as defined by Eurocode 2. (a) Stress trajectories, (b) transverse stresses, (c) strut and tie model, (d) idealised forces one side and (e) position of rebars.

tie model used in Eurocode 2 is shown in Figure 9.9c and considering moment equilibrium about X in Figure 9.9d

$$0.25H \times T = \left(\frac{0.5H + 0.65L_b}{4} - \frac{L_b}{4}\right) \times \frac{F}{2}$$

which rearranges to

$$T = \left(\frac{0.5H + 0.65L_b}{4} - \frac{L_b}{4}\right) \times \frac{2F}{H}$$

$$T = \frac{F}{4}\left(1 - \frac{0.7L_b}{H}\right) \tag{9.2}$$

9.3 DESIGN OF THE TIES

The strength of a tie is calculated using Equation 7.4

$$T = 0.87A_s f_{yk} \tag{9.3}$$

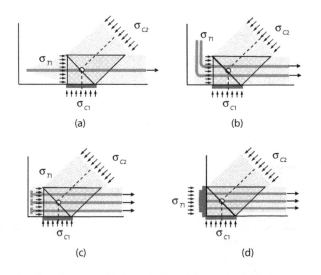

Figure 9.10 Examples of end anchorages. (a) Straight bar, (b) standard 90-degree hook, (c) T-headed bar and (d) bearing plate.

More than one rebar can be represented as a single tie, in which case the rebars are spread equally either side of the centre line of the notional tie in the STM. Particular care must be taken to ensure that ties are adequately anchored beyond node points; Figure 9.10 shows examples of end anchorages.

9.4 CONTROL OF COMPRESSION STRESSES

Compression struts are classified into two main types: prismatic and bottle shaped (see Figure 9.11). The important difference between these is that stress is spreading in bottle-shaped struts, whereas it is uniform for prismatic ones. This is important, because spreading stresses create sideways tensile stresses and these (a) cause tension cracking and (b) reduce compression strength. The sloping (diagonal) struts shown in Figure 9.6 are bottle shaped, whereas the short vertical struts located under loads are prismatic.

The struts and ties intersect at nodes, which fall into the three types shown in Figure 9.12. The allowable stresses depend on the type of node, with stresses falling in

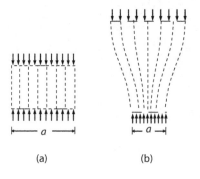

Figure 9.11 Strut compressive stress fields. (a) Prismatic and (b) bottle-shaped.

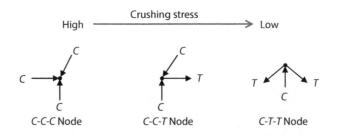

Figure 9.12 Node types and the strut crushing stresses.

the presence of tensile stresses from ties. The Eurocode 2 limiting stresses of the struts are as follows:

$$\text{Prismatic struts in CCC nodes, } \sigma_{Rd} = 1.00 \times v' f_{cd} \tag{9.4}$$

$$\text{Prismatic struts in CCT nodes, } \sigma_{Rd} = 0.85 \times v' f_{cd} \tag{9.5}$$

$$\text{Prismatic struts in CTT nodes, } \sigma_{Rd} = 0.75 \times v' f_{cd} \tag{9.6}$$

$$\text{Unreinforced bottle-shaped struts, } \sigma_{Rd} = 0.6 v' f_{cd} \tag{9.7}$$

$$\text{Reinforced bottle-shaped struts, } \sigma_{Rd} = 1.0 v' f_{cd} \tag{9.8}$$

where f_{cd} is the design compressive stress for concrete, given by Equation 7.1, and

$$v' = 1 - \frac{f_{ck}}{250} \tag{9.9}$$

Combining Equations 9.7, 9.9 and 7.1 provides the limit for unreinforced bottle-shaped struts, which tend to be critical over the other stress limits, i.e.,

$$\sigma_{Rd} = 0.6 \times \left(1 - \frac{f_{ck}}{250}\right) \times 0.567 f_{ck} \tag{9.10}$$

9.4.1 Reinforced bottle-shaped struts

If transverse (bursting) reinforcement is required, a submodel of the bottle-shaped strut is needed (see Figure 9.13). The dimensions of the bottle-shaped strut will depend on the width-to-length ratio of the strut (b/H). If b/H > 0.5, then the full discontinuity approach should be used, as shown in Figure 9.9c.

Providing reinforcement perpendicular to the strut angle is not usually practical; therefore, the bursting stress reinforcement should be provided in two layers: horizontal and vertical (as shown in Figure 9.14b). The combined tensile strength of the horizontal and vertical reinforcement is

$$T = 0.87 f_{yk} A_h \sin\theta + 0.87 f_{yk} A_v \cos\theta \tag{9.11}$$

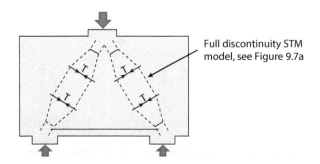

Figure 9.13 Strut and tie model for a bottle-shaped strut.

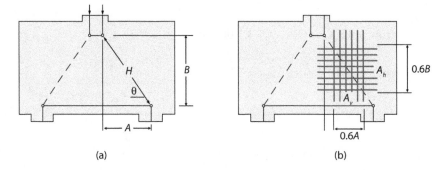

(a)   (b)

Figure 9.14 Bottle-shaped strut reinforcement. (a) Basic STM and (b) minimum spread of reinforcement.

where
   T is given by Equation 9.2 (H is shown in Figure 9.14a).
   A_h and A_v are the areas of reinforcement in the horizontal/vertical direction
   θ is the angle of the strut (see Figure 9.14a).

More useful are the rebar areas per metre width, which are given by

$$A_v \text{ per m width} = \frac{A_v}{0.3A} \tag{9.12}$$

$$A_h \text{ per m width} = \frac{A_h}{0.3B} \tag{9.13}$$

where A and B are shown in Figure 9.14a. These areas can be restricted to the central 0.6 region of the strut shown in Figure 9.14b or more sensibly extended to the whole of the diagonal strut for extra control of cracking.

9.4.2 The calculation of strut widths

The width of the diagonal struts needs to be calculated in order to determine the stress. The process for calculating strut width is illustrated in Figure 9.15a, which shows a Compression Compression Compression (CCC) node where the bottle-shaped strut width is

$$w_{C3} = L_b \sin\theta + w_{C2} \cos\theta \tag{9.14}$$

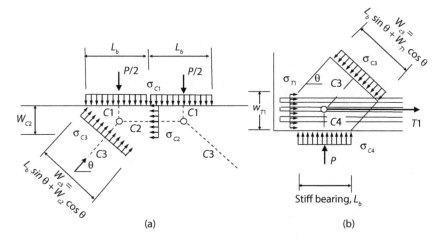

Figure 9.15 Calculation of strut width for the beam shown in Figure 9.14. (a) CCC Node and (b) CCT Node.

where w_{C2} is twice the depth to the compression strut $C1$, and L_b is the stiff bearing width, which in this case is half the width of the bearing plate. In the example shown in Figure 9.15a, the load is split into two separate loads; therefore, L_b is half the plate width. For the Compression Compression Tension (CCT) node shown in Figure 9.15b, the bottle-shaped strut width is

$$w_{C3} = L_b \sin\theta + w_{T1} \cos\theta \qquad\qquad (9.15)$$

where w_{T1} would be twice the depth to the centre of the rebar (see Figure 9.15).

9.5 MINIMUM REINFORCEMENT

Each face of a deep beam should have a minimum area of reinforcement to control cracking. This would usually be taken as 0.1% of the area of the concrete on each face, although for deep beams (>1 m) the minimum area is 0.2% on each face, provided horizontally and vertically to create a mesh. If this minimum area is greater than the area required from the STM calculations, then this replaces the STM reinforcement.

Example 9.1: Control of cracking for a column supporting a concentrated load

A 600 mm square column is subjected to an 840 kN concentrated load, applied through a 240 mm square bearing plate. Design the reinforcement at the head of the column to control cracking. $f_{yk} = 500$ N/mm² and $f_{ck} = 40$ N/mm².

Step 1: Formulate the strut and tie model. The D-region is taken as a square 600×600 mm region from the top of the column, as shown in Figure 9.16a. The remaining column is B-region; therefore, this is a partially discontinuous model (see section 9.2.1). The node

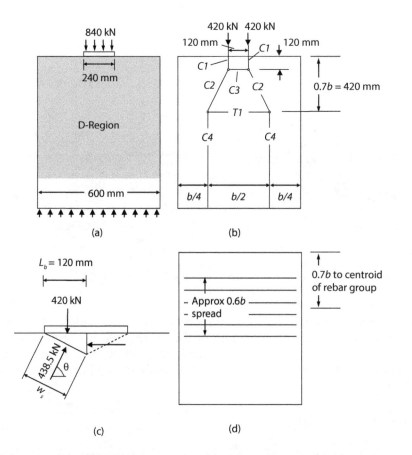

Figure 9.16 Square column STM. (a) Cross-section, (b) e strut and tie model, (c) calculation of ws for CCC node and (d) rebar distribution for T1.

positions can be defined using the approach illustrated in Figure 9.8c, where b = 600 mm and the depth to the centroid of the CCC node is 0.2b = 120 mm (see Figure 9.16b). This is a 2D model of a 3D problem, because splitting can occur on all four faces of the column. Therefore, the rebar (T1) is required in both orthogonal directions.

Step 2: Design the tie. Since this is a partial discontinuity, Equation 9.1 is applied

$$T = \frac{F}{4}\left(\frac{b - L_b}{b}\right)$$

$$T = \frac{840}{4}\left(\frac{600 - 240}{600}\right) = 126 \text{ kN}$$

And from Equation 9.3, the minimum cross-sectional area of rebar is

$$A_s = \frac{126 \times 10^3}{0.87 \times 500} = 290 \text{ mm}^2$$

Table 9.1 Sectional areas of groups of rebars, in mm²

Diameter (mm)	Number of bars									
	1	2	3	4	5	6	7	8	9	10
6	28	57	85	113	141	170	198	226	254	283
8	50	101	151	201	251	302	352	402	452	503
10	79	157	236	314	393	471	550	628	707	785
12	113	226	339	452	565	679	792	905	1018	1131
16	201	402	603	804	1005	1206	1407	1608	1810	2011
20	314	628	942	1257	1571	1885	2199	2513	2827	3142
25	491	982	1473	1963	2454	2945	3436	3927	4418	4909
32	804	1608	2413	3217	4021	4825	5630	6434	7238	8042
40	1257	2513	3770	5027	6283	7540	8796	10053	11310	12566

Since the reinforcement is hoop shaped, each rebar counts twice. Thus, 6 × 6 mm diameter rebar would provide an area of 340 mm² (i.e., 2 × 170 mm², see Table 9.1). The centroid of the group of bars needs to be 0.7b (420 mm) from the top (see Figure 9.16d). The six bars should be spread across the width of the tie member, which in this case is 0.6b or 360 mm (see Figure 9.16d). Therefore, 6 mm diameter shear links at 75 mm spacing would be sufficient. For extra control of cracking, this reinforcement should be continued to the top of the column. *Provide 6 mm diameter shear links at 75 mm spacing across the top 600 mm of the column.*

Step 3: Check compression stresses. The width of the bottle-shaped strut needs to be determined using Equation 9.14. The stiff bearing length (L_b) is 120 mm, because the model shown in Figure 9.16b splits the load into two equal 420 kN forces. The width of C3 is twice the distance from the top of the column to the centre of the node, i.e., $w_{C3} = 2 \times 120 = 240$ mm, and from the geometry of the model the angle of inclination of the diagonal strut is

$$\theta = \tan^{-1}\left(\frac{0.5 \times 600}{600/4 - 240/4}\right) = 73.3°$$

From equilibrium, the force in the bottle-shaped strut, C2, is

$$C2 = \frac{420}{\sin 73.3} = 438.5 \text{ kN}$$

From Equation 9.14

$$w_{C2} = L_b \sin\theta + w_{C3} \cos\theta$$

$$w_{C2} = 120 \sin(73.3) + 240 \cos(73.3) = 184 \text{ mm}$$

The compression stress is

$$\sigma_{C2} = \frac{438.5 \times 10^3}{184 \times 240} = 9.9 \text{ N/mm}^2$$

From Equation 9.10, the limiting stress is

$$\sigma_{Rd} = 0.60 \times \left(1 - \frac{40}{250}\right) \times 0.567 \times 40 = 11.4 \text{ N/mm}^2$$

Since this is greater than the applied stress, the strut is adequate. The 3D effects were not accounted for in this 2D estimate of σ_{c2}. A (more complicated) 3D model would have produced lower stresses.

Example 9.2: Deep beam

A deep beam supports a 400 × 400 mm square column with a force of 2600 kN and is supported by two 350 mm square columns (see Figure 9.17). Use the STM method to design the reinforcement. The material properties are f_{yk} = 500 N/mm² and f_{ck} = 40 N/mm².

Step 1: Formulate the strut and tie model. Since the span-to-depth ratio < 3, the entire deep beam can be considered as a D-region. Considering equilibrium and symmetry, the reactions are established and a STM is created, as shown in Figure 9.17b. In this case, a 1:2 slope on the diagonal struts works well and is simple, although other slopes are possible, i.e., you do not need to use a 1:2 slope.

Step 2: Design the ties. The force in the bottom tension member is

$$T_{AC} = \frac{1300}{\tan(63.4)} = 650 \text{ kN}$$

And from Equation 9.3

$$A_s = \frac{650 \times 10^3}{0.87 \times 500} = 1495 \text{ mm}^2$$

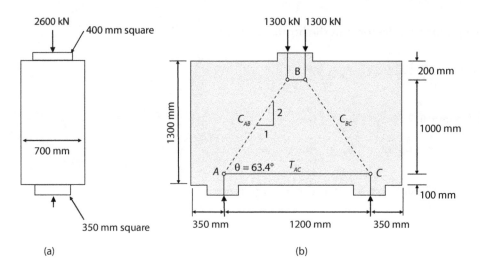

Figure 9.17 Deep beam. (a) Side view and (b) front view.

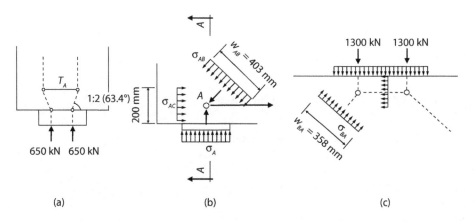

Figure 9.18 Node details. (a) Section A–A showing the transversal model of node A, (b) node A from the front and (c) node B from the front.

Provide 8 No. 16 mm diameter bars (1608 mm²).

A transverse strut and tie model is also needed to control cracking above the piles (see Figure 9.18a). The slope of the diagonal struts is assumed conservatively to be 1:2; therefore, the tie force is

$$T_A = \frac{1300}{4} = 325 \text{ kN}$$

And from Equation 9.3

$$A_s = \frac{325 \times 10^3}{0.87 \times 500} = 747 \text{ mm}^2$$

Provide four No. 16 mm diameter bars (say 804 mm²) as transversal reinforcement over the piles and double this below the supported column since the load is doubled.

Step 3: Check the compression stresses. From Equation 9.10, the limiting stress is

$$\sigma_{Rd} = 0.60 \times \left(1 - \frac{40}{250}\right) \times 0.567 \times 40 = 11.4 \text{ N/mm}^2$$

The strut force is

$$C_{AB} = \frac{1300}{\sin(63.4)} = 1453.9 \text{ kN}$$

From Equation 9.15, the width of the bottle-shaped strut AB at Node A is

$$w_{AB} = L_b \sin\theta + w_{AC} \cos\theta$$

$$w_{AB} = 350 \times \sin(63.4°) + 200 \times \cos(63.4°) = 403 \text{ mm}$$

And the compression stress is

$$\sigma_{AB} = \frac{1453.9 \times 10^3}{403 \times 350} = 10.3 \text{ N/mm}^2$$

From Equation 9.14, the width of the strut AB at Node B is

$$w_{BA} = \frac{400}{2} \times \sin(63.4°) + 400 \times \cos(63.4°) = 358 \text{ mm}$$

And the applied compression stress is

$$\sigma_{BA} = \frac{1453.9 \times 10^3}{358 \times 400} = 10.2 \text{ N/mm}^2$$

Since the applied stresses (10.3 and 10.2 N/mm²) are less than 11.4 N/mm², the strut need not be reinforced.

The limiting stress for prismatic struts in CCT nodes from Equation 9.5 is

$$\sigma_{Rd} = 0.85 \times \left(1 - \frac{40}{250}\right) \times 0.567 \times 40 = 16.2 \text{ N/mm}^2$$

The stress in the prismatic strut directly below Node A is

$$\sigma_A = \frac{1300 \times 10^3}{350 \times 350} = 10.6 \text{ N/mm}^2$$

And the stress induced by the rebar is

$$\sigma_{AC} = \frac{650 \times 10^3}{200 \times 350} = 9.3 \text{ N/mm}^2$$

Both of these stresses are significantly lower than 16.2 N/mm² and are not therefore critical. Indeed, prismatic strut stresses are rarely critical if bottle-shaped struts are unreinforced.

Step 4: Design the minimum reinforcement A mesh of orthogonal reinforcement with a minimum area of 0.2% of the cross section should be provided on all faces, because the beam depth exceeds 1 m. This area can most easily be calculated by considering the beam as a 700 mm deep slab and working out the area requirement per square metre using Table 9.2.

$$\text{Area} = \frac{0.2}{100} \times 1000 \times 700 = 1400 \text{ mm}^2/\text{m}$$

From Table 9.2, it can be seen that 16 mm diameter rebars at 125 mm centres provide 1608 mm²/m and are therefore adequate to be provided horizontally and vertically on all faces.

Table 9.2 Rebar cross-sectional area per metre width

Diameter	Spacing of bars in mm								
	50	75	100	125	150	175	200	250	300
6	565	377	283	226	188	162	141	113	94
8	1005	670	503	402	335	287	251	201	168
10	1571	1047	785	628	524	449	393	314	262
12	2262	1508	1131	905	754	646	565	452	377
16	4021	2681	2011	1608	1340	1149	1005	804	670
20	6283	4189	3142	2513	2094	1795	1571	1257	1047
25	9817	6545	4909	3927	3272	2805	2454	1963	1636

Example 9.3: Deep beam

A 450 mm wide beam supports a 450 mm square column and is supported on two 500 mm × 450 mm rectangular columns as shown in Figure 9.19. The ultimate applied load is 3360 kN. Develop a STM and calculate the reinforcement required. Material properties are f_{yk} = 500 N/mm² and f_{ck} = 35 N/mm².

Step 1: Formulate the strut and tie model. As the span-to-depth ratio is less than 3, the entire deep beam can be considered as a D-region. The centroid of the tension steel is assumed to be 100 mm above the soffit and all the diagonal struts have a 1:2 slope for simplicity. Using these assumptions, a STM is created as shown in Figure 9.20. In this example, the 1:2 slope works perfectly, although other angles can be used if the 1:2 slope does not fit the geometry.

Step 2: Design the ties. Considering member AC, from Equation 9.3

$$A_s = \frac{1120 \times 10^3}{500 \times 0.87} = 2575 \text{ mm}^2$$

Provide six No. 25 mm diameter rebars from Nodes A to E with the bar ends bent 90° at the ends of the beam to provide anchorage at the nodes. The vertical tie CD has the same

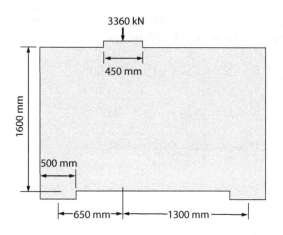

Figure 9.19 Side elevation of a 450 mm wide transfer beam.

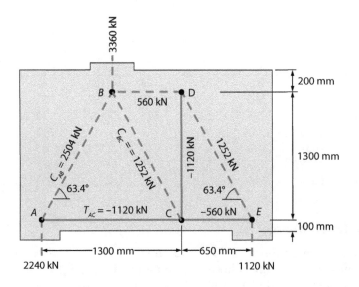

Figure 9.20 Forces in the struts and ties.

force and thus requires the same reinforcement. This should be spread across the tie width, which is 650 mm; therefore,

$$A_s/\mathrm{m}^2 = \frac{2575}{0.65} = 3962 \text{ mm}^2/\text{m}$$

From Table 9.2, it can be seen that 16 mm diameter links at 100 mm spacing will provide 2 × 2011 = 4021 mm²/m (the links pass through each side of the beam and therefore are counted twice in this area calculation).

Step 3: Check compression stresses. The length of stiff bearing under the central column must be calculated. The force $C_{BA} = 2 \times C_{BC}$; therefore, the 450 mm wide column is divided up on a 2/3 and 1/3 basis for the purpose of calculating L_b, i.e., the length of stiff bearing for strut BA is 300 mm and for BC is 150 mm. This arrangement gives an even distribution of stress under the supported column and is therefore consistent with the real condition. From Equation 9.14, the width of the strut BA at Node B is

$$w_{BA} = 300 \times \sin(63.4°) + 400 \times \cos(63.4°) = 447 \text{ mm}$$

From Equation 9.15, the width of strut AB at Node A is

$$w_{AB} = 500 \times \sin(63.4°) + 200 \times \cos(63.4°) = 537 \text{ mm}$$

$w_{BA} < w_{AB}$; therefore, the maximum applied stress in C_{AB} is at Node B, where

$$\sigma_{BA} = \frac{2504 \times 10^3}{447 \times 450} = 12.5 \text{ N/mm}^2$$

From Equation 9.10, the limiting stress in the bottle-shaped struts is

$$\sigma_{Rd} = 0.60 \times \left(1 - \frac{35}{250}\right) \times 0.567 \times 35 = 10.2 \text{N/mm}^2 < 12.5 \text{ N/mm}^2 \therefore FAIL$$

From Equation 9.8, the limiting stress for a reinforced strut is

$$\sigma_{Rd} = 1.00 \times \left(1 - \frac{35}{250}\right) \times 0.567 \times 35 = 17.1 \text{ N/mm}^2 > 12.5 \text{ N/mm}^2 \therefore PASS$$

The applied stress (12.5 N/mm²) is greater than the limit for unreinforced struts (10.2 N/mm²) but less than the limit for reinforced struts (17.1 N/mm²); therefore, the strut will be acceptable if reinforced. Now checking the stress in strut BC at Node B

$$w_{BC} = 150 \times \sin(63.4°) + 400 \times \cos(63.4°) = 313 \text{ mm}$$

$$\sigma_{BC} = \frac{1252 \times 10^3}{313 \times 450} = 8.9 \text{ N/mm}^2 < 10.2 \text{ N/mm}^2 \therefore PASS$$

Since strut AB is to be reinforced, the stresses are increased and the prismatic struts need to be checked. Node B is a CCC node, and from Equation 9.4 the limiting stress for a intersecting prismatic strut is

$$\sigma_{Rd} = 1.00 \times \left(1 - \frac{35}{250}\right) \times 0.567 \times 35 = 17.1 \text{ N/mm}^2$$

The stress in the prismatic strut directly above Node B is

$$\sigma_B = \frac{3360 \times 10^3}{450 \times 450} = 16.6 \text{ N/mm}^2 < 17.1 \text{ N/mm}^2 \therefore PASS$$

Node A is a CCT node; therefore, from Equation 9.5 the limiting stress is

$$\sigma_{Rd} = 0.85 \times \left(1 - \frac{35}{250}\right) \times 0.567 \times 35 = 14.5 \text{ N/mm}^2$$

And the applied stress below Node A is

$$\sigma_A = \frac{2240 \times 10^3}{500 \times 450} = 9.95 \text{ N/mm}^2 < 14.5 \text{ N/mm}^2 \therefore PASS$$

And the sideways stress induced by the tie (T_{AC}) is

$$\sigma_{AC} = \frac{1120 \times 10^3}{200 \times 450} = 12.4 \text{ N/mm}^2 < 14.5 \text{ N/mm}^2 \therefore PASS$$

Step 4: Design of the bursting reinforcement. In summary, bursting reinforcement is needed for strut AB but not for the other bottle-shaped struts (BC and DE). Using the design parameters for a strut with full discontinuity (Figure 9.9c) the length of stiff bearing $L_b = w_{BA} = 447$ mm and the strut length is

$$H = \sqrt{650^2 + 1300^2} = 1453 \text{ mm}$$

From Equation 9.2, the tension force illustrated in Figure 9.13 is

$$T = \frac{F}{4}\left(1 - \frac{0.7L_b}{H}\right)$$

L_b is taken as w_{BA}; therefore,

$$T = \frac{2504}{4}\left(1 - \frac{0.7 \times 447}{1453}\right) = 491 \text{ kN}$$

Equation 9.11 is

$$T = 0.87 f_{yk} A_h \sin\theta + 0.87 f_{yk} A_v \cos\theta$$

Letting $A_H = 2A_V$, this rearranges to

$$A_h = \frac{T}{0.87 f_{yk}(\sin\theta + 0.5 \times \cos\theta)}$$

$$A_h = \frac{491 \times 10^3}{500 \times 0.87(\sin 63.4 + 0.5 \times \cos 63.4)} = 1010 \text{ mm}^2$$

From Equation 9.13

$$A_h/m = \frac{1010}{0.3 \times 1.3} = 2590 \text{ mm}^2/m$$

Spread over the central $0.6 \times 1300 = 780$ mm (see Figure 9.14), and since $A_H = 2A_V$ from Equation 9.12

$$A_v/m = \frac{1010/2}{0.3 \times 0.65} = 2590 \text{ mm}^2/m$$

Spread over the central $0.6 \times 650 = 390$ mm. Inspection of Table 9.2 reveals that 16 mm diameter links at 150 mm centres provide $2 \times 1340 = 2680$ mm²/m of reinforcement (i.e. 1340 mm² on each face).

Step 5: Minimum reinforcement. In deep beams, the minimum area of rebar is 0.2% of the cross section; thus,

$$A_{min} = \frac{0.2}{100} \times 1000 \times 450 = 900 \text{ mm}^2/\text{m}$$

From Table 9.2, it can be seen that 16 mm rebar at 200 mm centres provided horizontally and vertically on all faces meets this requirement. This does not have to be provided in addition to the already calculated reinforcement, and the arrangement of rebar is summarised in Figure 9.21.

Example 9.4: Corbel

A 600 × 600 mm square column supports a girder through a corbel, illustrated in Figure 9.22. The ULS reaction from the girder is 750 kN applied through a 200 × 400 mm steel plate. Use the strut and tie method to determine the reinforcement required to support this load. The corbel is 600 mm wide.

Basic data f_{yk} = 500 N/mm², f_{ck} = 40 N/mm², 25 mm cover

Step 1: Formulate the strut and tie model. The node positions are shown in Figure 9.23. Initially, Node B was located 50 mm from the front face of the column. This created high stresses below Node B, which was moved to 100 mm from the column face as shown. The strut and tie model shown is statically determinate, so taking moments and resolving was used to determine the member forces. Figure 9.23a shows a 2D model through the corbel, although there are 3D effects. These are taken into consideration using the 2D model in the orthogonal plane shown in Figure 9.23b.

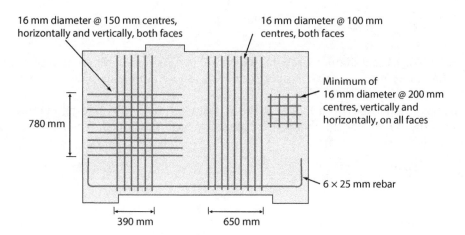

16 mm diameter @ 150 mm centres, horizontally and vertically, both faces

16 mm diameter @ 100 mm centres, both faces

Minimum of 16 mm diameter @ 200 mm centres, vertically and horizontally, on all faces

780 mm

6 × 25 mm rebar

390 mm   650 mm

Figure 9.21 Final arrangement of reinforcement.

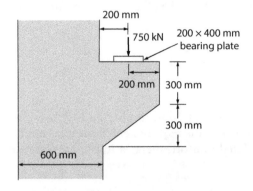

Figure 9.22 Corbel geometry and loads.

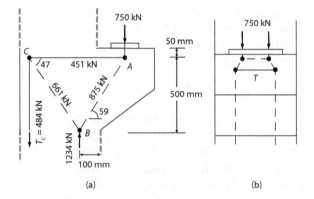

Figure 9.23 Strut and tie models for a corbel. (a) side elevation and (b) front elevation.

Step 2: Design the ties. The amount of reinforcement required in the ties is calculated by taking the higher of the two tensile forces at Node C. From Equation 9.3

$$A_s = \frac{T_C}{0.87f_{yk}} = \frac{484 \times 1000}{0.87 \times 500} = 1113 \text{ mm}^2$$

Provide 10 No. 12 mm diameter rebars, which Table 9.1 shows have an area of 1131 mm². These should be continuous past Nodes A and C as shown in Figure 9.24.

A transversal model shown in Figure 9.23b is assumed under the loading plate to prevent splitting of the corbel. For simplicity, a 2:1 load dispersal is assumed, which provides a splitting force of

$$T = \frac{375\,\text{kN}}{2} = 187.5 \text{ kN}$$

And from Equation 9.3

$$A_s = \frac{187.5 \times 1000}{0.87 \times 500} = 431.0 \text{ mm}^2$$

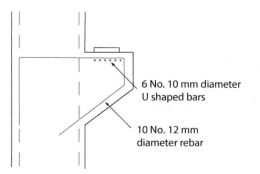

6 No. 10 mm diameter
U shaped bars

10 No. 12 mm
diameter rebar

Figure 9.24 Cross section showing rebar from strut and tie models (other rebar not shown).

Provide six No. 10 mm diameter rebars. In order to provide anchorage, these should be U-shaped; the rebar are shown in Figure 9.24.

Step 3: Check compression stresses. The stresses in the struts and nodes are checked and any unsatisfactory nodes are revised. Strut BA is critical over BC because of the higher force. From Equation 9.10, the limiting stress in the bottle-shaped strut is

$$\sigma_{Rd} = 0.6 \times \left(1 - \frac{40}{250}\right) \times 0.567 \times 40 = 11.4 \text{ N/mm}^2$$

And from Equation 9.15, Node A:

$$w_{AB} = 200 \times \sin(59) + 100 \times \cos(59) = 222.9 \text{ mm}$$

$$\sigma_{AB} = \frac{875 \times 10^3}{222.9 \times 400} = 9.8 \text{ N/mm}^2 \ngtr 11.4 \therefore PASS$$

And Node B:

$$w_{BA} \cong L_b \sin\theta = 200\sin(59) = 171 \text{ mm}$$

$$\sigma_{BA} = \frac{875 \times 10^3}{171 \times 600} = 8.53 \text{ N/mm}^2 \ngtr \sigma_{Rd} \therefore PASS$$

The bottle-shaped struts are unreinforced; therefore, it is not necessary to check the node stresses.

Example 9.5: Bridge pier

A bridge pier supports two 2800 kN loads placed on either side. A cross section (Figure 9.25) of the pier shows it is 8 m wide at the top and tapers to 6 m wide below the top face. The pier is 1 m deep. The loads are transferred to the pier through 700 × 350 mm square bearing plates, placed at 200 mm from either edge of the pier. Using the strut and tie method, design

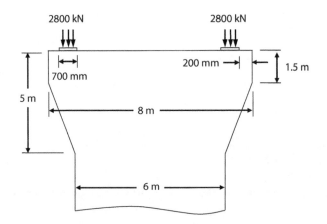

Figure 9.25 Bridge pier problem.

the reinforcement required at the top of the pier to prevent splitting cracks from occurring due to the concentrated loads, assuming $f_{yk} = 500$ N/mm^2 and $f_{ck} = 40$ N/mm^2.

Step 1: Formulate the strut and tie model. The whole bridge pier can be considered a D-region from the front face (see Figure 9.26a). The depth to the centre of the tie T is taken as 200 mm. This is the position that the centroid of the tension steel must be placed in. The diagonal strut is assumed to be inclined at a slope of 1:2 (63.4°), which produces a force of

$$C = \frac{2800}{\sin 63.4} = 3131.0 \text{ kN}$$

It follows from equilibrium of the node that a tensile force of 1400 kN is developed along the top of the pier. A transverse model is required to prevent splitting under the bearings (see Figure 9.26b). The dimensions of this are taken from the partial discontinuity member shown in Figure 9.8.

Step 2: Design ties. From Equation 9.3, the minimum area of steel along the top of the pier is

$$A_s = \frac{1400 \times 10^3}{0.87 \times 500} = 3220 \text{ mm}^2$$

Provide seven No. 25 mm diameter rebars, which from Table 9.1 provide 3436 mm^2.

A transversal model is used to prevent lateral cracks (Figure 9.26b). Since this is a partial discontinuity, Equation 9.1 is applied

$$T = \frac{F}{4}\left(\frac{b - L_b}{b}\right)$$

$$T_2 = \frac{2800}{4}\left(\frac{1000 - 350}{1000}\right) = 455 \text{ kN}$$

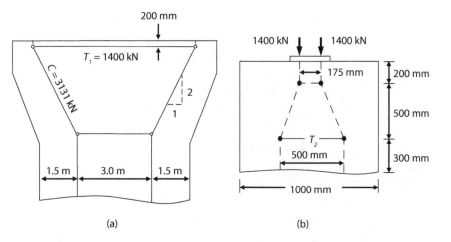

Figure 9.26 Bridge pier strut and tie models. (a) Main STM and (b) transverse STM.

And from Equation 9.3, the minimum cross-sectional area of rebar is

$$A_s = \frac{455 \times 10^3}{0.87 \times 500} = 1046 \text{ mm}^2$$

If the reinforcement under the bearings is hoop shaped, then each rebar counts twice. Thus, eleven No. 8 mm diameter rebars would provide an area of 1106 mm² (see Table 9.1). The centroid of the group of bars needs to be 0.7b (700 mm) from the top and spread across the width of the tie, which in this case is 0.6b or 600 mm (see Figure 9.26b). Therefore, 8 mm diameter links at 50 mm spacing would be more than sufficient. For extra control of cracking, this should be continued to the top of the pier. Thus, provide twenty No. 8 mm diameter links at 50 mm spacing under the bearings (see Figure 9.27).

Step 3: Check compression stresses. From Equation 9.10, the limiting stress is

$$\sigma_{Rd} = 0.60 \times \left(1 - \frac{40}{250}\right) \times 0.567 \times 40 = 11.4 \text{ N/mm}^2$$

The width of the diagonal strut next to the bearing from Equation 9.15 is

$$w_C = 700 \times \sin(63.43°) + 400 \times \cos(63.43°) = 805 \text{ mm}$$

And the applied stress is

$$\sigma_{Ed} = \frac{3131 \times 10^3}{805 \times 350} = 11.1 \text{ N/mm}^2$$

Since this is less than σ_{Rd}, this bottle-shaped strut passes. Since the strut is unreinforced, it is not necessary to check the prismatic strut stresses.

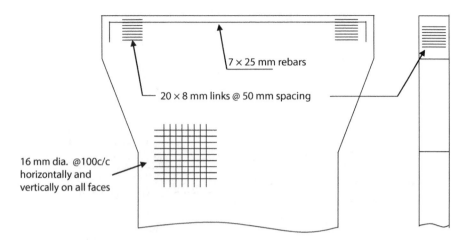

Figure 9.27 Rebar arrangement for bridge pier.

Step 4: Design the minimum reinforcement. The minimum reinforcement horizontally and vertically on all faces is 0.2% on both faces, because this deep beam has a depth > 1.0 m. Therefore:

$$A_s = \frac{0.2}{100} \times 1000 \times 1000 = 2000 \text{ mm}^2 / \text{m}$$

From Table 9.2, 16 mm diameter rebar at 100 mm centres satisfies this requirement, providing 2011 mm²/m.

Example 9.6: Bridge diaphragm

The solid diaphragm section in a box girder bridge transfers the web forces from the hollow box sections to the columns (see Figure 9.28). Use the strut and tie method to calculate the reinforcement required in the solid diaphragm section to prevent cracking due to the 7500 kN column reaction. The geometry and design loads are shown in Figure 9.29.

Basic data

f_{yk} = 500 N/mm² and f_{ck} = 35 N/mm².

Step 1: Formulate the strut and tie model. The whole diaphragm is a D-region and the strut and tie model selected to resist the support reaction is shown in Figure 9.30. A transverse STM over the support is not required, because the lower part of the diaphragm is a high compression zone due to the hogging moments and the prestress force.

The forces can be calculated from equilibrium of the nodes:

$$C_{12} = \frac{3750}{\sin(44.2)} = 5379 \text{ kN}$$

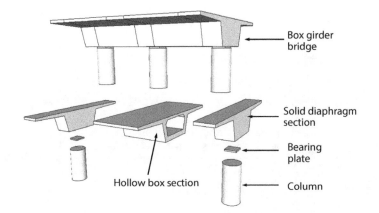

Figure 9.28 Prestressed concrete box-girder bridge.

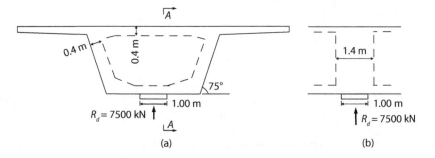

Figure 9.29 Cross section with geometry (in metres) and loads. (a) Cross section and (b) section A–A.

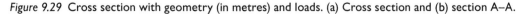

$$C_{11} = \frac{3750}{\tan(44.2)} = 3856 \text{ kN}$$

$$T_{23} = \frac{3750}{\cos(15)} = 3882 \text{ kN}$$

$$T_{22} = 5379 \times \cos(44.2) - 3882 \times \sin(15) = 2852 \text{ kN}$$

Step 2: Design the ties. From Equation 9.3, the area of the top reinforcement to resist, T_{22}, is

$$A_{s22} = \frac{2852 \times 10^3}{0.87 \times 500} = 6556 \text{ mm}^2$$

From Table 9.1, provide a 22×20 mm diameter rebar, which provides 6912 mm².

And the area of the shear reinforcement to resist, T_{23}, is

$$A_{s23} = \frac{3882 \times 10^3}{0.87 \times 500} = 8924 \text{ mm}^2$$

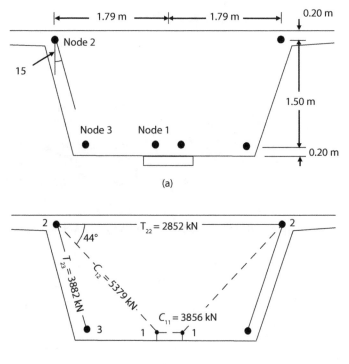

Figure 9.30 Strut and tie model for the diaphragm. (a) Node positions and (b) forces.

Provide 23 × 16 mm diameter shear links, which provide 2 × 201 × 23 = 9246 mm² (see Table 9.1). The arrangement is sketched in Figure 9.31.

Step 3: Check the compression stresses. From Equation 9.10, the limiting stress for an unreinforced bottle-shaped strut is

$$\sigma_{Rd} = 0.60 \times \left(1 - \frac{35}{250}\right) \times 0.567 \times 35 = 10.24 \text{ N/mm}^2$$

At Node 1: From Equation 9.14, the width of the bottle-shaped strut between Nodes 1 and 2 is

$$w_{12} = L_b \sin\theta + w_{C2}\cos\theta = 500 \times \sin(44°) + 400 \times \cos(44°) = 635 \text{ mm}$$

And the applied stress is

$$\sigma_{12} = \frac{5379 \times 10^3}{635 \times 1000} = 8.5 \text{ N/mm}^2$$

Since this is less than 10.24 N/mm², this is a pass.

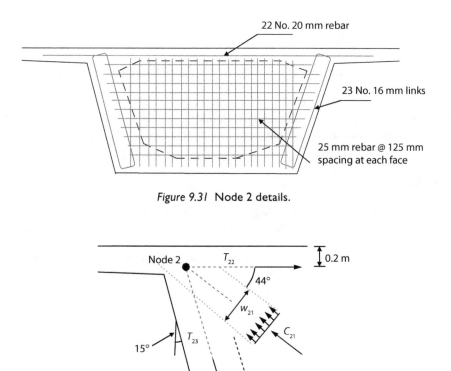

Figure 9.31 Node 2 details.

Figure 9.32 Arrangement of reinforcement, cross section.

At Node 2: The calculation of the width of the bottle-shaped strut (C_{21}) at Node 2 is described in Figure 9.32. The calculation is complicated by the inclination of the tie T_{23}, although Equation 9.15 can still be applied. The length of stiff bearing is approximately

$$L_b = \text{web width} \times \cos(15) = 400\cos(15) = 386 \text{ mm}$$

And

$$w_{21} = 386 \times \sin(44°) + 400 \times \cos(44°) = 555 \text{ mm}$$

And the applied stress is

$$\sigma_{21} = \frac{5379 \times 10^3}{573 \times 1400} = 6.7 \text{ N/mm}^2$$

Since this is less than 10.24 N/mm², this is a pass and from inspection the CTT node stresses due to forces T_{22} and T_{23} will not be critical.

Step 4: Minimum reinforcement. The arrangement of reinforcement is established, as shown in Figure 9.31. Each face of this deep beam should have a minimum area of reinforcement to prevent cracking. Since the depth >1 m deep, the minimum area is 0.2% on each face, provided horizontally and vertically.

$$A_s = \frac{0.2}{100} \times 1000 \times 1400 = 2800 \text{ mm}^2/\text{m} \Rightarrow H25@175 \text{ mm c/c}$$

From Table 9.2, it can be seen that 25 mm diameter rebar at 175 mm centres will be sufficient and the final rebar arrangement is summarised in Figure 9.32.

REFERENCES

FIB, 2011. *Design examples for strut-and-tie models*. Technical Report No. 61.

Goodchild, C.H., Morrison, J. and Vollum, R., 2014. *Strut and tie models - How to design concrete structures using Eurocode 2*. London: MPA The Concrete Centre.

MacGregor, J.G. and Wight, J.K., 2011. *Reinforced concrete: Mechanics and design*. 6th Edition. Pearson.

Chapter 10

Control of cracking in reinforced concrete

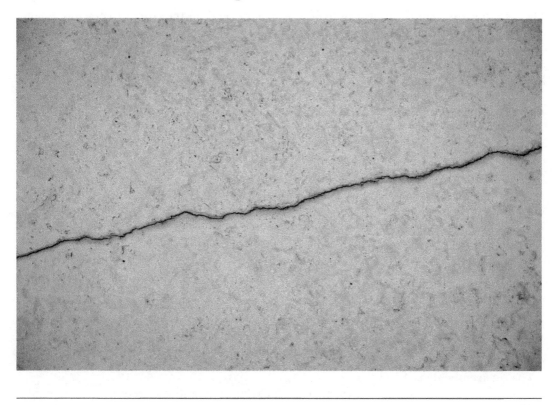

Cracking is normal in reinforced concrete. If they are less than approximately 0.5 mm wide, cracks will not usually cause corrosion, which is normally caused by the permeation of dissolved salt or carbon dioxide through the cover. Cracking needs to be controlled, mainly because it is unsightly and causes complaints.

Cracking occurs when the tensile strain exceeds the cracking strain. Tensile strains occur due to loads, shrinkage and temperature changes. This chapter is concerned with all these actions.

The chemical reaction between cement and water is exothermic and causes an effect known as the *heat of hydration*. When the concrete cools, the resulting contraction is the cause of the first cracks to form, known as *early age cracking*. This can lead to internal cracks in deep sections, because of the temperature gradient due to differential rates of cooling between the inner and outer parts. Early age cracks widen over time due to climatic temperature changes and drying, and these effects combine to form a life cycle of movements, i.e.,

Stage 1 Heat of hydration movements → within the first 3–4 days
Stage 2 Winter–summer thermal movements → every 6 months
Stage 3 Shrinkage due to drying → potentially up to 10 years

These effects build up with time, and cracks may take years to develop.

10.1 HEAT OF HYDRATION SHRINKAGE

The temperature rise due to heat of hydration will depend on the size of the member, with deep sections developing the highest temperatures. In extreme situations, the temperature rise can be 60°C or more. Cracking results from the temperature gradient within a member or restraint to contractions during cooling. The insulation properties of formwork influence heat build-up, as does the type of cement used. Insulated formwork can be helpful in reducing internal cracking, even though the temperature will increase. This is because insulation can reduce the internal temperature gradient.

Cements incorporating large percentages of ground granulated blast furnace slag or pulverised fuel ash develop lower heats of hydration than 100% ordinary Portland cement concretes, and the codes of practice contain methods to estimate the temperature rises for different types of cements.

The heat of hydration expansion does not induce compression stresses, because the concrete is not hardened. Problems occur when the concrete gains strength and cools. The temperature change ΔT shown in Figure 10.1 causes a contraction and the corresponding tensile strain, ε, is

$$\varepsilon = \alpha \times \Delta T \tag{10.1}$$

And the corresponding movement, δ, is

$$\delta = \varepsilon \times L \tag{10.2}$$

where
L is the member length.
α is the coefficient of linear expansion, which for steel and concrete is approximately 12×10^{-6} per °C.

If the length L is replaced with crack spacing (S), then Equation 10.2 can be rewritten to provide an approximate prediction of crack width, w

$$w = \varepsilon \times S \tag{10.3}$$

10.2 DRYING SHRINKAGE

Drying shrinkage is due to loss of water from (a) evaporation and (b) water being chemically bonded during the hydration of the cement, known as *autogenous shrinkage*. These shrinkage strains widen the cracks formed due to heat of hydration movements (early age cracks).

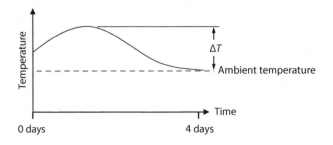

Figure 10.1 Heat of hydration temperature changes.

Codes of practice present procedures for estimating shrinkage strains, which depend on factors such as the rate of drying, which is influenced by the atmospheric humidity (i.e. inside or outside) and the size of the member. Rates of drying therefore decrease with increasing member size, thereby reducing shrinkage strains.

10.3 CREEP STRAIN

Long-term stress permanently deforms concrete with time in a process known as *creep*. This can be beneficial because it can reduce cracking due to shrinkage; however, creep does increase deflections because creep strains are significantly greater than the initial elastic strains (see Figure 10.2). Therefore, long-term deflection calculations need to be based on modified values of Young's modulus.

Codes of practice contain detailed procedures for calculating short-term Young's modulus values and the modification factor to account for creep, which depends on the duration of loading and the age at loading, since fresh concrete creeps more than mature concrete. The strength grade and aggregate type affect Young's modulus, and creep strains increase with increased moisture content. Therefore, the creep strains are also linked with member size and atmospheric humidity, since these factors affect the rate of drying and moisture content.

10.4 CRACKING DUE TO RESTRAINED SHRINKAGE

Many elements, such as walls, are not subjected to significant loads. Instead, the main loading is from shrinkage movements which induce tensile stresses. These stresses are transferred to the steel when the concrete cracks. For example, Figure 10.3 shows a wall cast onto a foundation. The wall is restrained from shrinking by the foundation, which was cast first. Since shrinkage is restrained, tension cracks can form, and if uncontrolled these can create visual defects that are particularly obvious after rain.

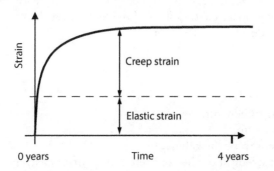

Figure 10.2 Strain under constant stress versus time.

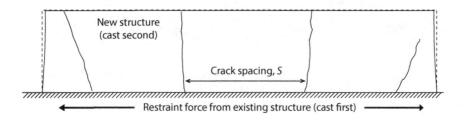

Figure 10.3 Cracking due to restrained shrinkage for a wall cast onto a slab.

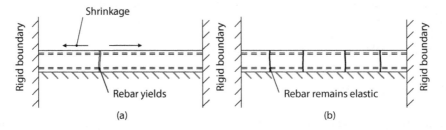

Figure 10.4 Cracking due to restrained shrinkage for a slab. (a) Insufficient rebar and (b) sufficient rebar.

This can be controlled by the adequate provision of rebar, which has the effect of reducing the crack spacing, as illustrated in Figure 10.4. Since crack width is a function of crack spacing, the crack width will also be reduced (see Equation 10.3). The maximum crack width is often limited to 0.2 mm, because cracks smaller than 2 cm self-seal with time.

If concrete is insufficiently reinforced, then the rebar will yield when the first crack forms as the element shrinks due to heat of hydration movement (see Figure 10.4a). Since the tensile strength of the rebar < tensile strength of the slab, all subsequent shrinkage will be taken up by yielding of the rebar in the first crack formed. However, if the tensile strength of the rebar > concrete, then the rebar in the cracks will remain elastic. Elastic movements are much smaller than plastic ones; therefore, the crack width will be less. More cracks will be formed to take up the movement (see Figure 10.4b), but the crack width will be less.

The guiding principle to control cracking is to ensure that the tensile strength of the rebar is not less than the tensile strength of the concrete at 3 days after casting.

Rebar strength ≥ Concrete tensile strength 3 days after casting

The crack pattern developed during the first few days is known as *early age cracking* and all subsequent shrinkage movements will be accommodated by widening of these cracks. The tensile strength at 3 days after casting is used in calculating the minimum area of reinforcement and the primary design requirement is

$$A_s f_{yk} \geq A_c f_{ct.3days}$$

where
A_s is the area of reinforcement.
A_c is the area of concrete.
f_{yk} is the tensile strength of the rebar.
$f_{ct,3days}$ is the tensile strength of the concrete 3 days after casting.

This rearranges to provide the minimum area of reinforcement

$$A_s = \frac{A_c f_{ct.3days}}{f_{yk}} \tag{10.4}$$

This would normally be distributed equally between both faces of a slab or wall. The 3-day strength of concrete is approximately half of the 28 day strength, and the tensile strength of concrete is approximately 1/10th of the crushing strength; therefore,

$$f_{ct.3days} = f_{ck}/20 \tag{10.5}$$

This helps to ensure that the steel remains elastic in the cracks. Importantly, it does not guarantee that crack widths will be within acceptable limits. Therefore, crack width calculations are also required.

10.5 CALCULATION OF CRACK WIDTHS

Estimation of crack width is based on simple empirically based methods that give very approximate solutions. As a simple guide, cracking is likely if the tensile strain exceeds $100\ \mu\varepsilon$ ($\times 10^{-6}$), in which case the crack width (w) is approximately

$$w = S \times \varepsilon_r \tag{10.6}$$

where
  S is the approximate crack spacing.
  ε_r is the restrained shrinkage strain.

The restrained shrinkage strain is approximately

$$\varepsilon_r = 0.75 \times \varepsilon_{\text{free}} R \tag{10.7}$$

where
  $\varepsilon_{\text{free}}$ is the strain that would occur if there was no restraint to shrinkage, known as the *free shrinkage strain*.
  R is the restraint factor.

The 0.75 factor accounts for creep strains relieving heat of hydration shrinkage strains. Creep is more significant for long-term drying shrinkage, where this factor can be reduced to 0.5, although 0.75 is assumed throughout this chapter for simplicity.

Restraint factor. The restraint factor depends on the degree of restraint to shrinkage. The worst case is when a wall or slab is cast between two rigid boundaries, in which case it would equal 1.0 (see Figure 10.5a). This is why slabs should *not* be cast in alternate bays, since this would provide restraint on two opposite edges, as illustrated in Figure 10.5b. This is also why the construction sequence is important for water-retaining structures.

Many slabs and walls are restrained from shrinking when cast onto an existing structure. For example, the wall shown in Figure 10.6 will tend to shrink after casting,

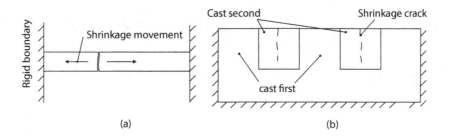

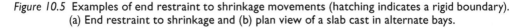

Figure 10.5 Examples of end restraint to shrinkage movements (hatching indicates a rigid boundary). (a) End restraint to shrinkage and (b) plan view of a slab cast in alternate bays.

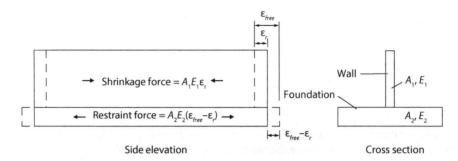

Figure 10.6 A wall is cast onto a slab, which provides a restraint to shrinkage.

although the shrinkage is restrained by the foundation, which was cast first. From force equilibrium

Shrinkage force = Restraint force

Looking at Figure 10.6, this becomes

$$A_1 E_1 \varepsilon_r = A_2 E_2 \left(\varepsilon_{\text{free}} - \varepsilon_r \right)$$

Dividing through by ε_r

$$A_1 E_1 \frac{\varepsilon_r}{\varepsilon_{\text{free}}} = A_2 E_2 \left(1 - \frac{\varepsilon_r}{\varepsilon_{\text{free}}} \right)$$

If

$$\varepsilon_r = R \times \varepsilon_{\text{free}}$$

This becomes

$$A_1 E_1 R = A_2 E_2 - R A_2 E_2$$

And the restraint factor is

$$R = \frac{A_2 E_2}{A_1 E_1 + A_2 E_2} \tag{10.8}$$

In this calculation, Young's modulus for the freshly cast wall would be approximately 70% that of the hardened foundation, i.e., $E_1 = 0.7 E_2$.

Crack spacing. The crack spacing can be estimated using the following expression:

$$S = 3.4c + 0.34 \, \varphi / \rho_p \tag{10.9}$$

where
    c is the cover.
    ρ_p is the reinforcement ratio.

The reinforcement ratio is the ratio of the rebar and concrete areas, i.e.,

$$\rho_p = \frac{A_s}{A_c} \qquad (10.10)$$

where
  A_s is the area of steel.
  A_c is the area of concrete.

Equation 10.9 indicates that rebar diameter should be minimised when controlling cracking.

Example 10.1: Control of cracking in a basement slab

The basement slab of a building is cast in one section and restrained against shrinkage by piled walls. The slab is 290 mm thick and 25 mm cover is provided, f_{ck} = 40 N/mm^2, f_{yk} = 500 N/mm^2.

1. Determine the maximum reinforcement spacing to control shrinkage cracking if 12 mm diameter rebars are used.
2. Heat of hydration causes the temperature to increase to 35°C before falling back to an ambient temperature of 15°C after 3 days. Estimate the crack width due to this fall in temperature.
3. A drying shrinkage strain of 80 με is expected over the 2 years after casting. Calculate the final crack width.

1. From Equation 10.5, the 3 day tensile strength is

$$f_{ct.3days} = f_{ck}/20 = 2.0 \text{ N/mm}^2$$

And from Equation 10.4, considering a 1000 mm wide section of the 290 mm thick slab

$$A_s = \frac{A_c f_{ct.3days}}{f_{yk}} = \frac{290 \times 1000 \times 2.0}{500} = 1160 \text{ mm}^2/\text{m}$$

This equates to 580 mm^2/m top and bottom and in both directions. Provide 12 mm diameter rebars at 175 mm centres in both directions, which equates to 1292 mm^2/m (see Table 9.2).

2. From Equation 10.10, the reinforcement ratio is

$$\rho_p = \frac{A_s}{A_c} = \frac{1292}{1000 \times 290} = 4.455 \times 10^{-3}$$

And from Equation 10.9, the crack spacing approximately is

$$S = 3.4c + 0.34 \ \varphi/\rho_p$$

$$S = 3.4 \times 25 + \frac{0.34 \times 12}{4.455 \times 10^{-3}} = 1001 \text{ mm}$$

From Equation 10.1, the free shrinkage strain is

$$\varepsilon_{\text{free}} = \alpha \times \Delta T = 12 \times 10^{-6} \times (35 - 15) = 240 \times 10^{-6}$$

The question says that the slab is prevented from shrinking by the piled walls that surround the slab. In the absence of other information, it is conservatively assumed that these are rigid boundaries as illustrated in Figure 10.5a, although in practice they will have some flexibility. Thus, the restraint factor, R, is 1.0 and from Equation 10.7 the restrained shrinkage strain is

$$\varepsilon_r = 0.75 \times \varepsilon_{\text{free}} R = 0.75 \times 240 \times 10^{-6} \times 1.0 = 180 \times 10^{-6}$$

And from Equation 10.6, the crack width is

$$w = S \times \varepsilon_r = 1001 \times 180 \times 10^{-6} = 0.18 \text{ mm}$$

Since this is less than 0.2 mm, the crack will probably self-seal over time.

3. The free shrinkage strain due to shrinkage after 2 years is 80 $\mu\varepsilon$.

$$\varepsilon_r = 0.75 \times \varepsilon_{\text{free}} R = 0.75 \times 80 \times 10^{-6} \times 1.0 = 60 \times 10^{-6}$$

$$w = S \times \varepsilon_r = 1001 \times 60 \times 10^{-6} = 0.06 \text{ mm}$$

Final crack width is

$$w = 0.18 + 0.06 = 0.24 \text{ mm}$$

Example 10.2: Control of cracking for a wall cast onto a foundation

A wall is cast onto a foundation as shown in Figure 10.7. The engineer is worried that the foundation will restrain shrinkage of the wall and lead to unsightly cracking. The wall does not resist significant loading and only requires reinforcement to control cracking. The cover is 25 mm thick, $f_{ck} = 30 \text{ N/mm}^2$, $f_{yk} = 500 \text{ N/mm}^2$.

1. Determine the minimum reinforcement spacing if 6 mm diameter rebars are used.
2. Determine the crack width due to early thermal cracking if the wall heats up by 15°C due to the heat of hydration.
3. Determine the crack width after 2 years if drying causes 120$\mu\varepsilon$ of shrinkage strain.

1. From Equation 10.5, the 3 day tensile strength is

$$f_{ct.3days} = f_{ck}/20 = 1.5 \text{ N/mm}^2$$

From Equation 10.4, the minimum area of reinforcement for a 1000 m wide section of wall is

$$A_s = \frac{A_c f_{ct.3days}}{f_{yk}} = \frac{175 \times 1000 \times 1.5}{500} = 525 \text{ mm}^2/\text{m}$$

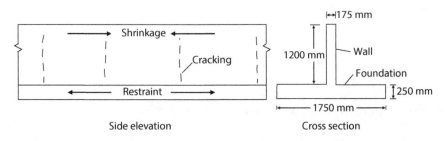

Figure 10.7 Wall restrained from shrinking after being cast onto a foundation.

Provide 6 mm diameter horizontal rebars at 100 mm centres on each face, which provide a total of 2 × 283 = 566 mm²/m (see Table 9.2). The vertical rebar can be designed using the minimum area for beams, which is less onerous (Equation 7.35).

2. From Equation 10.10, the reinforcement ratio is

$$\rho_p = \frac{A_s}{A_c} = \frac{566}{1000 \times 175} = 3.234 \times 10^{-3}$$

And from Equation 10.9, the crack spacing is

$$S = 3.4c + 0.34\,\varphi/\rho_p$$

$$S = 3.4 \times 25 + \frac{0.34 \times 6}{3.234 \times 10^{-3}} = 716 \text{ mm}$$

And from Equation 10.1, the free shrinkage strain due to a 15°C temperature change is

$$\varepsilon_{\text{free}} = \alpha \times \Delta T = 12 \times 10^{-6} \times 15 = 180 \times 10^{-6}$$

It is assumed that Young's modulus for the freshly cast concrete is only 70% of the more mature foundation; therefore,

$$E_1 = 0.70 \times E_2$$

And from Equation 10.8, the restraint factor is

$$R = \frac{A_2 E_2}{0.7 A_1 E_2 + A_2 E_2}$$

$$R = \frac{250 \times 1750}{0.7 \times 175 \times 1200 + 250 \times 1750} = 0.75$$

And from Equation 10.7, the restrained strain is

$$\varepsilon_r = 0.75 \times R \times \varepsilon_{\text{free}} = 0.75 \times 0.75 \times 180 \times 10^{-6} = 101 \times 10^{-6}$$

From Equation 10.6, the crack width is

$$w = \varepsilon \times S = 101 \times 10^{-6} \times 716 = 0.07 \text{ mm}$$

This is well within acceptable limits.

3. Drying shrinkage causes the early age cracks to widen by

$$w = \varepsilon \times S = 120 \times 10^{-6} \times 716 = 0.09 \text{ mm}$$

Therefore, the crack width after 2 years is 0.07 + 0.09 = 0.16 mm.

10.6 CALCULATION OF CRACK WIDTHS FOR BEAMS

For beams, the crack width (w) is approximately equal to the crack spacing (S) multiplied by the strain in the steel (ε_s), i.e.,

$$w = \varepsilon_s \times S \tag{10.11}$$

The strain in the steel can be easily calculated from basic mechanics because beams should be elastic under SLS loads. The crack spacing (S) can be estimated using the following empirical expression for beams and slabs

$$s = 3.4c + 0.17 \ \phi/\rho_{p,\text{ eff}} \tag{10.12}$$

where
c is the cover.
ϕ is the rebar diameter.
$\rho_{p,\text{ eff}}$ is the effective reinforcement ratio given by

$$\rho_{p,\text{ eff}} = \frac{A_s}{A_{ct}} \tag{10.13}$$

where
A_s is the area of steel within the tension zone.
A_{ct} is the area of the tension zone.

The tension zone is the hatched region shown in Figure 10.8 for beams and slabs.

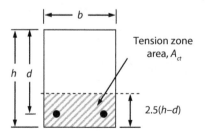

Figure 10.8 The 'tension zone' area shown as the hatched region for members in bending.

Example 10.3: Calculation of crack width in a beam

Figure 10.9a shows a beam subjected to a 30 kN.m moment under SLS loads. Estimate the crack width if the rebar diameter is 20 mm, E_c is 18,000 N/mm², E_s is 210,000 N/mm² 25 mm is the cover.

The concrete in compression will have a linear stress distribution, because the beam remains elastic under SLS loads (see Figure 10.9c). The position of the neutral axis must be determined. To do this, the method of transformed sections is used to convert the rebar into an equivalent area of concrete, as illustrated in Figure 10.9b, i.e.,

$$\text{Transformed area of rebar} = A_s \frac{E_s}{E_c} = 628 \times \frac{210000}{18000} = 7327 \text{ mm}^2$$

It is necessary to determine the distance from the top of the member to the neutral axis, shown as x in Figure 10.9a. This is also the centroid of area of the transformed section, which is located by taking moments of area about the top face, i.e.,

$$200x \times \frac{x}{2} + 7327 \times 310 = (200x + 7327)x$$

$$100x^2 + 7327x - 2271370 = 0$$

The roots are $x = 118.5$ mm and -191.7 mm. The centroid of the concrete compression force is centred at $x/3$ below the top of the beam (see Figure 10.9c). Taking moments about this point

$$M = \sigma_s A_s \left(d - \frac{x}{3} \right)$$

Rearranging provides the stress in the steel

$$\sigma_s = \frac{M}{A_s \left(d - \dfrac{x}{3} \right)}$$

$$\sigma_s = \frac{30 \times 10^6}{628 \left(310 - \dfrac{118.5}{3} \right)} = 176.6 \text{ N/mm}^2$$

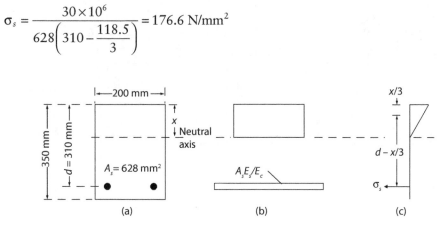

Figure 10.9 Calculation of crack width. (a) Cross-section, (b) 'transformed section' and (c) stress distribution.

And the corresponding strain is

$$\varepsilon_s = \frac{\sigma_s}{E_s} = \frac{176.6}{210000} = 841 \times 10^{-6}$$

The crack spacing must now be determined, and from Figure 10.8 the tension zone depth is

$$h_c = 2.5(h - d) = 2.5(350 - 310) = 100 \text{ mm}$$

From Equation 10.13

$$\rho_{p,\,eff} = \frac{A_s}{A_{ct}} = \frac{628}{200 \times 100} = 0.0314$$

From Equation 10.12, the crack spacing is

$$s = 3.4c + 0.17\phi/\rho_{p,\,eff} = 3.4 \times 25 + 0.17 \times 20 / 0.0314 = 193 \text{ mm}$$

And finally from Equation 10.11, the crack width is

$$w = \varepsilon_s \times S = 841 \times 10^{-6} \times 193 = 0.16 \text{ mm}$$

10.7 CONTROL OF CRACKING DUE TO SOLAR GAIN

Solar gain is a temperature rise due to sunshine and the resulting expansions are a cause of cracking. For example, in buildings, this can result in cracking of columns. Consider the building shown in Figure 10.10. By combining Equations 10.1 and 10.2, the end movement due to an increase in roof temperature (ΔT) is

$$\delta = \alpha \times \Delta T \times L$$

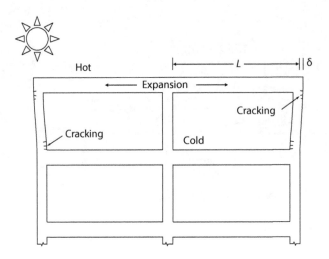

Figure 10.10 Building cross section showing column cracks due to solar gain.

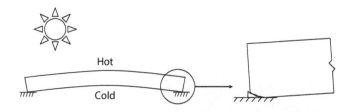

Figure 10.11 Dishing induced by a temperature gradient in a beam.

If α = 12 $\mu\varepsilon$/°C, ΔT = 30°C and L = 50 m, then the end movement would be

$$\delta = 12 \times 10^{-6} \times 30 \times 50 \times 10^{3} = 18 \text{ mm}$$

This can cause cracking, for example, 2 cracking of columns as shown in Figure 10.10.

Solar gain can also result in a strain gradient within a member. This induces curvature causing 'dishing', which results in support movements and in extreme cases to cracking, as illustrated in Figure 10.11.

Example 10.4: Movements in a roof slab due to solar gain

The roof of a hotel heats up and expands, causing the edge columns to move sideways by 20 mm, as illustrated in Figure 10.12a. The storey to storey height (L) is 4.0 m and the columns are 400 mm square, with 25 mm cover; each is reinforced with four No. 25 mm diameter rebars + 12 mm links. The columns are far less stiff (EI) than the beams; therefore, the supports can be assumed to be clamped. Estimate the crack width.

The first step is to estimate the strain in the rebar. An approximate solution is to idealise the column as fixed supports subjected to a sideways movement of Δ, as illustrated in Figure 10.12b. Formulae tables will tell us that the end moments are

$$M = \frac{6EI\Delta}{L^2}$$

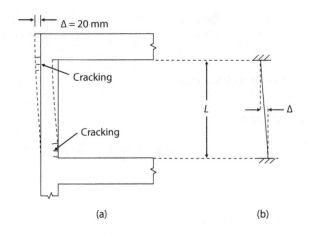

(a)

(b)

Figure 10.12 Analysis of column cracking due to expansion of roof slab. (a) Cross-section and (b) idealisation of deflection.

The engineer's beam equation tells us that

$$M = \frac{\sigma I}{y}$$

Combining these

$$\frac{6EI\Delta}{L^2} = \frac{I\sigma}{y}$$

Since

$$E = \frac{\sigma}{\varepsilon}$$

The above rearrange to provide the strain as

$$\varepsilon = \frac{6\Delta y}{L^2}$$

If y is half the depth of the column, the outer edge strain is

$$\varepsilon = \frac{6 \times 20 \times 200}{4000^2} = 1500 \times 10^{-6}$$

The next step is to calculate area of the tension zone (see Figure 10.8). The effective depth is

$$d = 400 - 25 - 12 - 25/2 = 350.5 \text{ mm}$$

And the area of the tension zone as shown in Figure 10.8 is

$$A_{ct} = 400 \times 2.5 \times (400 - 350.5) = 49500 \text{ mm}^2$$

A total of four rebars are used in the column (one in each corner), although only two are in the tension zone. Therefore, the area of two No. 25 mm diameter rebars = 981 mm² and from Equation 10.12

$$\rho_{p,\text{ eff}} = A_s/A_{ct} = 981/49500 = 0.0198$$

Equation 10.12, the crack spacing is

$$s = 3.4c + 0.17\phi/\rho_{p,\text{ eff}} = 3.4 \times 25 + 0.17 \times 25/0.0198 = 300 \text{ mm}$$

Equation 10.11, the crack width is

$$w = \varepsilon_s \times S = 1500 \times 10^{-6} \times 300 = 0.45 \text{ mm}$$

This approximate analysis indicates that cracking may be a problem.

Example 10.5: Dishing of a floor slab

The roof of a shopping mall car park comprises 450 mm deep precast concrete floor slabs, which span 16 m between supports. A temperature gradient of 30°C is induced between the top and bottom of the slab due to solar gain.

1. Estimate the dishing deflection and support movements.
2. Describe the problems that would likely occur if the precast units were not supported using appropriate bearings.

Basic data $\alpha = 12\ \mu\varepsilon/°C$ and $E_c = 20,000\ N/mm^2$

1. The total change in temperature between the top and bottom is 30°C, although the temperature difference between the top and centre of slab is half of this, at 15°C. From Equation 10.1, the resulting strain at the outer edge of the slab is

$$\varepsilon = \alpha\Delta T = 12 \times 10^{-6} \times 15 = 180 \times 10^{-6}$$

The engineer's beam equation tells us that

$$\frac{M}{I} = \frac{\sigma}{y} = \frac{E}{R} \tag{10.14}$$

Since $E = \sigma/\varepsilon$, the curvature $(1/R)$ is

$$\frac{1}{R} = \frac{\varepsilon}{y}$$

Inputting the strain induced by the solar gain

$$\frac{1}{R} = \frac{180 \times 10^{-6}}{450/2} = 0.8 \times 10^{-6}$$

This curvature would be induced along the full length of the beam and would produce a strain distribution the same as a beam subjected to equal and opposite end moments, as illustrated in Figure 10.13a. In that case, standard tables of formulae tell us that the end slope and midspan deflection are

$$\theta = \frac{ML}{2EI} \tag{10.15}$$

And

$$\Delta = \frac{ML^2}{8EI} \tag{10.16}$$

Equation 10.14 rearranges to

$$\frac{1}{R} = \frac{M}{EI} \tag{10.17}$$

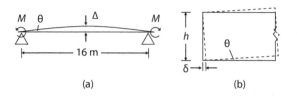

Figure 10.13 Beam idealisation for dishing movements. (a) Beam idealisation and (b) support movements.

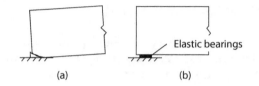

Figure 10.14 Dishing of a slab due to solar gain. (a) Resulting cracking and (b) solution.

Combining Equations 10.15 and 10.17 gives the end slope

$$\theta = \frac{L}{2} \times \frac{1}{R} = \frac{16000}{2} \times 0.8 \times 10^{-6} = 6.4 \times 10^{-3} \text{ radians}$$

And combining Equations 10.16 and 10.17 gives the midspan deflection

$$\Delta = \frac{L^2}{8} \times \frac{1}{R} = \frac{16000^2}{8} \times 0.8 \times 10^{-6} = 25.6 \text{ mm}$$

Assuming that the point of end-rotation occurs at the top corner of the slab (Figure 10.13b), the support sideways movement is approximately

$$\delta = \theta \times h = 6.4 \times 10^{-3} \times 450 = 3 \text{ mm}$$

2. The frictional force resulting from 3 mm of movement would be sufficient to cause cracking of the concrete supports, as shown in Figure 10.14a. Problems can be solved with elastic bearings, as illustrated in Figure 10.14b, because dishing does not cause stress unless movements are restrained.

Problems

Solutions to these problems are provided at https://www.crcpress.com/9781498741217

P.10.1. A slab is restrained against shrinkage by stiff perimeter walls. The slab is 200 mm thick, 25 mm cover is provided, f_{ck} = 40 N/mm² and E_c = 18,000 N/mm².
   a. Determine the maximum reinforcement spacing if 10 mm diameter rebars are used to control cracking.
   b. Heat of hydration causes the temperature to increase to 20°C before falling back to an ambient temperature of 10°C after 3 days. Estimate the crack width due to this fall in temperature.

c. The slab was cast in the summer, but the temperature falls to −2°C during the extreme winter cold. Recalculate the crack width.

d. A further 120 με of shrinkage is expected over a 2-year period after casting. Recalculate the maximum crack width.

Ans. (a) $A_{s.min}$ = 800 mm²/m use 10 mm diameter at 175 mm c/c top and bottom, (b) s = 842 mm and w = 0.076 mm, (c) 0.167 mm and (d) 0.243 mm.

P.10.2. Precast concrete floor slabs that span 12 m between supports and are 300 mm deep. Determine the vertical displacement and horizontal end-rotation at the supports due to dishing of the slabs caused by a temperature gradient of 35°C existing between the top and bottom of the slab resulting from heating by the summer sun. The coefficient of thermal expansion is 12 με/°C, E_c = 20,000 N/mm², f_{ck} = 40 N/mm².

Ans. Δ = 25.2 mm, θ = 8.4 × 10⁻³ radians, δ = 2.52 mm.

REFERENCES

Alexander, S.J., 2014. *Design for movements in buildings*. London: CIRIA.

Concrete Society, 2008. *Movement, restraint and cracking in concrete structures*. Camberley: Concrete Society. Publication No. TR67.

Chapter 11

Timber beams, columns and trusses

Timber is a non-ductile material, which means that it fails in a brittle manner when the failure stress is exceeded (see Figure 11.1). For this reason, plastic design is not allowed and all strength calculations are based on elastic principles. This makes the analysis relatively simple, based on straightforward mechanics.

This chapter does not dwell on how to determine the design values of material properties, which can be easily determined using one of the many codes of practice. Instead, first principles are used to calculate the strength of beams and columns, as well as the strengthening of timber sections with steel plates and finally the design of trusses. Lateral torsional buckling is not considered, although it can occur in deep sections if the compression half of the member is free to buckle sideways. Fortunately, most timber sections have the compression half-restrained by the floor or roof that they support, so this is not a common problem.

11.1 MATERIAL PROPERTIES

Engineers have the choice to either specify a certain strength grade of timber or a species, although in practice strength grades are most popular. The codes of practice help by specifying design stresses, known as *grade stresses* for various grades and species. Grade stresses include shear, bending, tension and compression stresses. The engineer can modify these to account for moisture content, load duration, load sharing, section size and notching.

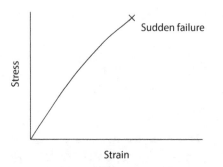

Figure 11.1 Stress versus strain for timber.

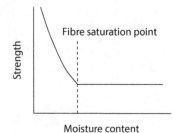

Figure 11.2 Effect of moisture content on strength.

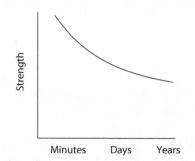

Figure 11.3 Effect of load duration on strength.

Moisture content. Grade stresses are specified for dry timber, because commercial timber is supplied dried as standard. However, the strength falls if timber is allowed to get wet or damp, as illustrated in Figure 11.2. Therefore, if timber is going to be exposed to damp or rain, then grade stress and Young's moduli are reduced using a modification factor.

Load duration. Timber is significantly stronger when loads are applied for a short duration, as illustrated in Figure 11.3. The grade stresses are modified depending on the load duration. Long-term loading normally means dead loads, whereas short-term loads include wind and snow. The shorter the duration, then the higher the modification factor.

Load sharing. Grades stresses and Young's modulus can be increased if closely spaced members jointly support a load, i.e., floor joists or rafters. This is because imperfections (such as knots) are less critical when more than one member supports a load.

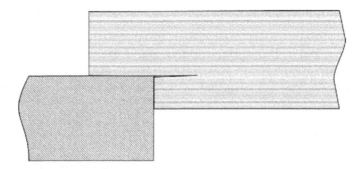

Figure 11.4 Splitting of a timber beam with a notched end.

Section size. The failure stresses increase slightly as the depth of timber sections decreases. For this reason, the codes include a modification factor to increase grade stresses for small timber sections.

Notching. Sharp changes in cross section are generally avoided in engineering, because they introduce stress concentrations that can induce cracking, especially in brittle materials like timber. Despite this, the ends of timber beams are often notched, as shown in Figure 11.4, in which case the shear stresses are modified accordingly.

11.2 SHEAR STRENGTH

The elastic shear stress equation gives the shear stress (τ) a distance y from the neutral axis

$$\tau = \frac{VA'\overline{y}}{b_o I} \tag{11.1}$$

where
    I is the second moment of area of the cross section.
    V is the applied shear force.
    A' is the area above the shear plane considered (see Figure 6.13).
    b_o is the width of the shear plane.
    $\overline{y}$ is the distance from the neutral axis to the centroid of the zone above the shear plane.

Shear flow. From a design perspective, shear flow (q) is more useful, where shear flow is

$$q = \tau \times b_o \tag{11.2}$$

Combining Equations 11.1 and 11.2

$$q = \frac{VA'\overline{y}}{I} \tag{11.3}$$

This can be used to determine the spacing of screws, because

$$q = \frac{\text{screw strength}}{\text{screw spacing}} \tag{11.4}$$

Screw spacing is measured along the length of the member and screw strength is the load capacity of the given screw. For example, screws may be spaced at 100 mm centres along the length of a beam and the code of practice may tell us that the shear strength of the screw is 700 N. This would produce a shear flow of 7 N/mm.

Rectangular cross sections. For a section of width b and depth h, the second moment of area is

$$I = \frac{bh^3}{12}$$ (11.5)

The maximum shear stress occurs at the midpoint where $A' = bh/2$ and $\bar{y} = h/4$, which when input into Equation 11.1 provide

$$\tau = \frac{V \times \dfrac{bh}{2} \times \dfrac{h}{4}}{b \times \dfrac{bh^3}{12}}$$

which simplifies to

$$\tau = \frac{3V}{2bh}$$ (11.6)

which rearranges to give the shear strength for an allowable shear stress

$$V = \frac{2}{3} \times bh\tau$$ (11.7)

11.3 BENDING STRENGTH

Since timber is brittle, plastic design is not allowed under any circumstances. Deep beams in which the compression half is unrestrained against lateral movement can suffer from *lateral torsional buckling*, although this check is rarely required for joists or rafters. If beams are strengthened using steel plates, then the *method of transformed sections* can be used to determine stresses and deflections.

The elastic bending strength is

$$M_{el} = \sigma_{Rd} W_{el}$$ (11.8)

where

f_d is the design stress in bending.

W_{el} is the elastic section modulus, which is the second moment of area divided by the distance from the centroid to the extreme fibre of the cross section.

For a rectangular section of width b and depth h

$$W_{el} = \frac{I}{y} = \frac{bh^3}{12} \times \frac{1}{h/2}$$

$$W_{el} = \frac{bh^2}{6}$$ (11.9)

Example 11.1: Floor beam

Rectangular timber beams 80 mm wide and 300 mm deep are used as floor beams (joists). The beams are restrained against lateral torsional buckling and span 4 m between simple supports. Determine the maximum uniformly distributed load if

1. Bending stresses are limited to 9.8 N/mm².
2. Shear stresses are limited to 1.34 N/mm².
3. Deflection is limited to 16 mm and Young's modulus is 8000 N/mm².

1. Consider bending stresses. From Equation 11.9, the elastic section modulus is

$$W_{el} = \frac{bh^2}{6} = \frac{80 \times 300^2}{6} = 1.2 \times 10^6 \, mm^3$$

From Equation 11.8

$$M_{el} = \sigma_{Rd} W_{el}$$

$$M_{el} = 9.8 \times 10^{-6} \times 1.2 \times 10^6 = 11.8 \, kN.m$$

If the applied moment reaches the elastic limit

$$M_{el} = \frac{wL^2}{8}$$

which rearranges to

$$w = \frac{8 \times 11.8}{4^2} = 5.9 \, kN/m$$

2. Consider shear stresses. From Equation 11.7

$$V = \frac{2}{3} \times bh\tau_{max}$$

$$V = \frac{2 \times 80 \times 300 \times 1.34}{3} \times 10^{-3} = 21.4 \, kN$$

The maximum shear force is

$$V = \frac{wL}{2}$$

If the applied shear force equals the shear strength

$$21.4 = \frac{w \times 4}{2}$$

which solves to w = 10.7 kN/m.

3. Consider deflection. From Equation 11.5

$$I = \frac{80 \times 300^3}{12} = 180 \times 10^6 \, mm^4$$

For a UDL the maximum deflection, ignoring shear deflection is

$$\Delta = \frac{5wL^4}{384EI} \tag{11.10}$$

which rearranges to

$$w = \frac{384EI}{5L^4}$$

$$w = \frac{384 \times 8000 \times 180 \times 10^6 \times 16}{5 \times 4000^4} = 6.9 \, N/mm = 6.9 \, kN/m$$

Example 11.2: Built-up member

A box beam is constructed using four timber sections screwed together as shown in Figure 11.5. If each screw has a shear capacity of 1300 N and they are spaced (in pairs, one each side) at 100 mm centres along the length of the beam, determine the shear force required to cause failure of the screws.

The second moment of area of the box section is

$$I = \frac{220 \times 300^3 - 140 \times 270^3}{12} = 265.4 \times 10^6 \, mm^4$$

From Equation 11.4, the shear flow is

$$q = \frac{\text{screw strength}}{\text{screw spacing}} = \frac{2 \times 1300}{100} = 26 \, N/mm$$

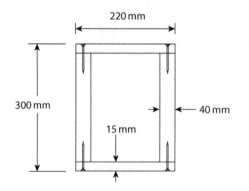

Figure 11.5 Cross section through beam built with screws.

At the screwed joint

$$\bar{y} = 300/2 - 15/2 = 142.5 \text{ mm}$$

And

$$A' = 220 \times 15 = 3300 \text{ mm}^2$$

Rearranging Equation 11.3

$$V = \frac{Iq}{A'\bar{y}}$$

$$V = \frac{265.4 \times 10^6 \times 26}{3300 \times 142.5} \times 10^{-3} = 14.7 \text{ kN}$$

Therefore, failure of the screws should occur at a shear force of 14.7 kN.

Example 11.3: Section strengthened with steel plates

A beam spans 6 m and is subjected to a point load at midspan of 18 kN, in addition to a UDL of 7.33 kN/m. The section is built up from a solid timber section with steel plates on the top and bottom faces (see Figure 11.6). Young's moduli are 8000 N/mm² for timber and 210,000 N/mm² for steel.

1. Determine the distribution of bending stresses at midspan.
2. If the plate is connected to the timber using rows of three screws, each with a shear capacity of 2 kN, determine the spacing of the rows of screws.
3. Determine the distance from the support that the spacing of the screws can be doubled (i.e. the 50% shear force point).
4. Determine the midspan deflection when supporting only a UDL of 5.5 kN/m.

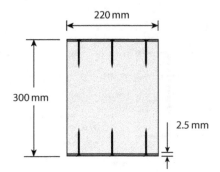

Figure 11.6 Cross section through a timber beam strengthened with steel plates screwed to the top and bottom faces.

1. The midspan moment is

$$M = \frac{PL}{4} + \frac{wL^2}{8} = \frac{18 \times 6}{4} + \frac{7.33 \times 6^2}{8} = 60 \text{ kN.m}$$

This is a composite beam, and therefore the *transformed sections method* is needed. The modular ratio is

$$n = \frac{E_1}{E_2} \qquad\qquad (11.11)$$

In this case, the timber will be transformed into steel; therefore,

$$n = \frac{8000}{210000} = 0.038$$

The timber is transformed into an equivalent width of steel as shown in Figure 11.7b

$$b' = n \times \text{timber width} = 0.038 \times 220 = 8.38 \text{ mm}$$

The second moment of area of the *transformed section* is

$$\frac{220 \times 300^3}{12} - \frac{(220 - 8.38) \times (300 - 5)^3}{12} = 42.3 \times 10^6 \text{ mm}^4$$

The stress distribution in the section is calculated using the engineer's beam equation, i.e.,

$$\sigma = \frac{My}{I}$$

The outer fibre stress is

$$\sigma = \frac{60 \times 10^6 \times (\pm)150}{42.3 \times 10^6} = \pm213 \text{ N/mm}^2$$

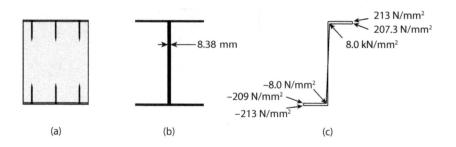

Figure 11.7 Bending stress distribution at midspan. (a) Strengthened beam, (b) transformed section and (c) bending stress distribution.

On the inside of the steel plate

$$\sigma = \frac{60 \times 10^6 \times \pm(150 - 2.5)}{42.3 \times 10^6} = \pm 209 \text{ N/mm}^2$$

The stresses at the top and bottom of the timber section are

$$\sigma = n \times \frac{My}{I} = 0.038 \times \frac{60 \times 10^6 \times \pm(150 - 2.5)}{42.3 \times 10^6} = \pm 8.0 \text{ N/mm}^2$$

And the stress distribution is sketched in Figure 11.7c.

2. The shear force at the supports is

$$V = \frac{P}{2} + \frac{wL}{2} = \frac{18}{2} + \frac{7.33 \times 6}{2} = 31 \text{kN}$$

At the screwed joint:

$$\bar{y} = 300 / 2 - 2.5 / 2 = 148.75 \text{ mm}$$

And

$$A' = 220 \times 2.5 = 550 \text{ mm}^2$$

From Equation 11.3

$$q = \frac{VA'\bar{y}}{I}$$

$$q = \frac{31 \times 10^3 \times 550 \times (150 - 0.5 \times 2.5)}{42.3 \times 10^6} = 60.0 \text{ N/mm}$$

From Equation 11.4 for the sets of three screws, each with a 2000 N capacity, the screw spacing is

$$\text{screw spacing} = \frac{\text{screw strength}}{q} = \frac{3 \times 2000}{60.0} = 100 \text{ mm}$$

3. The shear force diminishes by 7.33 kN for every metre away from the support. If the 50% shear force point is z metres from the support, then

$$31 - z \times 7.33 = \frac{31}{2}$$

which solves to $z = 2.1$ m or, in other words, the screw spacing can be doubled 2.1 m away from the supports.

4. From Equation 11.10, the midspan deflection when subjected to a UDL of 5.5 kN/m (=5.5 N/mm) is

$$\Delta = \frac{5 \times 5.5 \times 6000^4}{384 \times 210000 \times 42.3 \times 10^6} = 10.5 \text{ mm}$$

Note that Young's modulus for steel was used in the above calculation. This is because the timber was transformed into steel when calculating I.

Example 11.4: Timber and steel sandwich beam (flitch beam)

A beam is made of a steel plate sandwiched between two wooden beams, as sketched in Figure 11.8. The beam spans 5 m between simple supports and supports a UDL of 1.4 kN/m dead and 1.0 kN/m imposed. Young's moduli are 8000 N/mm² and 210,000 N/mm² for the timber and steel, respectively.

1. Determine the midspan deflection.
2. Determine the stresses in the timber and steel under factored loads.

1. The second moment of areas of the timber and steel are

$$I_t = 2 \times \frac{75 \times 200^3}{12} = 100 \times 10^6 \text{ mm}^4$$

$$I_s = \frac{10 \times 175^3}{12} = 4.47 \times 10^6 \text{ mm}^4$$

The stiffness (EI) of the timber and steel are

$$E_t I_t = 8000 \times 100 \times 10^6 = 800 \times 10^9 \text{ N.mm}^2$$

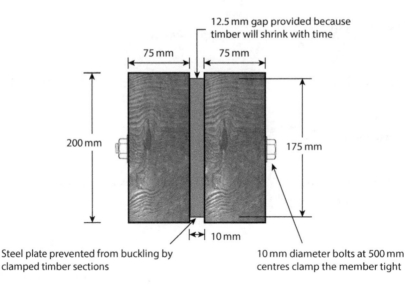

Figure 11.8 Cross section through a flitch beam (connecting bolts not shown).

$$E_s I_s = 210000 \times 4.47 \times 10^6 = 938 \times 10^9 \text{ N.mm}^2$$

And the total stiffness of the compound section is

$$\sum EI = (800 + 938) \times 10^9 = 1738 \times 10^9 \text{ N.mm}^2$$

The SLS load is

$$w_{sls} = 1.4 + 1.0 = 2.4 \text{ kN/m}$$

And the midspan deflection is

$$\Delta = \frac{5wL^4}{384\Sigma EI}$$

$$\Delta = \frac{5 \times 2.4 \times 5000^4}{384 \times 1738 \times 10^9} = 11 \text{ mm}$$

2. The ULS load is

$$w_{uls} = 1.35 \times 1.4 + 1.5 \times 1.0 = 3.39 \text{ kN/m}$$

The midspan moment is

$$M = \frac{wL^2}{8} = \frac{3.39 \times 5^2}{8} = 10.6 \text{ kN.m}$$

The moment is shared between the timber and steel sections. The proportion each section carries is dependent on the relative stiffness.

The moment supported by the timber is

$$M_t = \frac{E_t I_t}{\Sigma EI} \times M = \frac{800}{1738} \times 10.6 = 4.9 \text{ kN.m}$$

And the moment in the steel is

$$M_s = \frac{E_s I_s}{\Sigma EI} \times M = \frac{938}{1738} \times 10.6 = 5.7 \text{ kN.m}$$

The top and bottom stresses in the timber are

$$\sigma_t = \frac{M_t(\pm)y}{I_t} = \frac{4.9 \times 10^6 \times (\pm)100}{100 \times 10^6} = \pm 4.9 \text{ N/mm}^2$$

And the steel stresses are

$$\sigma_s = \frac{M_s(\pm)y}{I_s} = \frac{5.7 \times 10^6 \times (\pm)87.5}{4.47 \times 10^6} = \pm 112 \text{ N/mm}^2$$

11.4 COMPRESSION STRENGTH

Members subjected to a combination of compression and bending need to be checked against the combined loading. The Gordon–Rankine method will produce a quick estimate of compression strength, although it tends to be conservative in comparison with code-based methods. Using this method, the axial buckling force is

$$N_{b,\text{Rd}} = \left(\frac{1}{N_{\text{crush}}} + \frac{1}{N_{\text{cr}}} \right)^{-1} \tag{11.12}$$

where
$N_{b,\text{Rd}}$ is the design buckling load in the absence of bending moments.
N_{cr} is the (Euler) elastic critical buckling force.
N_{crush} is the crushing capacity of the member.

$$N_{\text{crush}} = \sigma_{\text{Rd}} \times \text{Area} \tag{11.13}$$

where σ_{Rd} is the design stress (in compression) and the elastic critical buckling force is

$$N_{\text{cr}} = \frac{\pi^2 EI}{L_{\text{cr}}^2} \tag{11.14}$$

where L_{cr} is the effective length, which is calculated in the same way as for any other compression member (see Chapter 3). The crushing capacity is equal to the cross-sectional area multiplied by the design stress, E is Young's modulus and I is the second moment of area of the member about the axis of buckling (usually the minor axis). Equation 11.12 works, because $N_{b,\text{Rd}} \to N_{\text{crush}}$ when slenderness is low and $N_{b,\text{Rd}} \to N_{\text{cr}}$ when slenderness is high.

11.5 COMPRESSION AND BENDING

Members subjected to compression and bending (beam columns) must satisfy the following interaction equation

$$\frac{N}{N_{b,\text{Rd}}} + \frac{\alpha_y M_y}{M_{\text{el},y}} + \frac{\alpha_z M_z}{M_{\text{el},z}} \leq 1 \tag{11.15}$$

where
N is the applied axial force.
M_y and M_z are applied moments about the major and minor axes, respectively.
$M_{\text{el},y}$ and $M_{\text{el},z}$ are the major and minor axis elastic moment capacities, respectively.

α_y and α_z are the moment amplification factors for the major and minor axes, respectively. These are calculated using the following approximate expressions:

$$\alpha_y = \frac{1}{1 - \dfrac{N}{N_{cr,\,y}}} \tag{11.16}$$

And

$$\alpha_z = \frac{1}{1 - \dfrac{N}{N_{cr,\,z}}} \tag{11.17}$$

where $N_{cr,y}$ and $N_{cr,z}$ are elastic critical values calculated using L_{cr} and I for their respective buckling axes. The amplification factor is $\alpha \rightarrow 1.0$ for low slenderness columns, although it rapidly becomes significant as column length increases, because Young's modulus is low for timber.

Equation 11.15 requires modification if lateral torsional buckling is an issue.

Example 11.5: Timber beam column

Timber struts 100 mm square in cross section are used as horizontal props. Each spans 2.5 m and the supports are effectively pinned. Young's modulus is 4000 N/mm², the design stress is limited to 10.5 N/mm² and the density is 400 kg/m³.

1. Determine the buckling force in the absence of moments.
2. Determine if each strut can resist a compression force of 20 kN (inclusive of load factors) in addition to the (factored) bending moment induced by the self-weight.
3. Determine what accidental sideways force a strut could resist without breaking.

1. The second moment of area and cross-sectional area are

$$A = 100^2 \text{ mm}^2$$

$$I = \frac{100 \times 100^3}{12} = 8.33 \times 10^6 \text{ mm}^4$$

The strut is pinned at the supports; therefore, $L_{cr} = 2500$ mm and from Equation 11.14 the elastic critical buckling force is

$$N_{cr} = \frac{\pi^2 EI}{L_{cr}^2}$$

$$N_{cr} = \frac{\pi^2 \times 4000 \times 8.33 \times 10^6}{2500^2} \times 10^{-3} = 52.6 \text{ kN}$$

And from Equation 11.13

$$N_{crush} = \sigma_{Rd} \times Area = 10.5 \times 100^2 \times 10^{-3} = 105 \text{ kN}$$

And using Equation 11.12, the compression strength is

$$N_{b, Rd} = \left(\frac{1}{N_{crush}} + \frac{1}{N_{cr}} \right)$$

$$N_{b, Rd} = \left(\frac{1}{52.6} + \frac{1}{105} \right)^{-1} = 35.0 \text{ kN}$$

2. The self-weight is

$$w = m \times g \times A = 400 \times 9.81 \times 10^{-3} \times 0.1^2 = 0.039 \text{ kN/m}$$

And the factored self-weight induced moment is

$$M = \frac{wL^2}{8} = \frac{1.35 \times 0.039 \times 2.5^2}{8} = 0.041 \text{ kN.m}$$

From Equation 11.16, the amplification of moments factor is

$$\alpha = \frac{1}{1 - N/N_{cr}} = \frac{1}{1 - 20/52.6} = 1.613$$

The elastic section modulus of the timber section from Equation 11.9 is

$$W_{el} = \frac{bh^2}{6} = \frac{100^3}{6}$$

From Equation 11.7, the elastic moment capacity is

$$M_{el} = 10.5 \times \frac{100^3}{6} \times 10^{-6} = 1.75 \text{ kN.m}$$

Now the interaction equation (Equation 11.15) is populated

$$\frac{N}{N_{b, Rd}} + \frac{\alpha_y M_y}{M_{el, y}} + \frac{\alpha_z M_z}{M_{el, z}} \leq 1$$

$$\frac{20}{35} + \frac{1.613 \times 0.041}{1.75} + \frac{1.613 \times 0}{1.75} = 0.609 \tag{11.18}$$

Since 0.609 < 1.0, this strut should be safe, although no allowance has been made for moments due to accidental forces.

3. Assuming the accidental force (P) is applied at midspan, the associated moment is

$$M = \frac{PL}{4} = 0.625P$$

Now adding this to Equation 11.18

$$\frac{20}{35} + \frac{1.613 \times 0.041}{1.75} + \frac{1.613 \times 0.625P}{1.75} = 1.0$$

which solves to $P = 0.68$ kN, which is equivalent to the weight of an average size person.

Example 11.6: Timber truss

A flat roof is supported by simply supported trusses, as sketched in Figure 11.9. These support a dead load (inclusive of self-weight) of 1.0 kN/m and an imposed load of 0.75 kN/m. The top and bottom chord members are made of 100 mm square timber sections with Young's modulus of 7200 N/mm² and a design stress of 7.0 N/mm².

1. Determine the maximum compression force in the top chord under ULS loads.
2. Determine the buckling capacity of the top chord if it is restrained against sideways movement by the roof decking.
3. Determine if the top chord can support the ULS loading.
4. Estimate the midspan deflection of the truss under SLS loading.

1. From Equation 1.3, the factored uniformly distributed design load is

$$w = 1.35g_k + 1.5q_k = 1.35 \times 1.0 + 1.5 \times 0.75 = 2.475 \text{ kN/m}$$

And the midspan moment is

$$M = \frac{wL^2}{8} = \frac{2.475 \times 10^2}{8} = 30.9 \text{ kN.m}$$

From Equation 3.13, taking moments about a chord provides the axial force in the other chord

$$N = \frac{M}{\text{lever arm}} = \frac{30.9}{1.0} = 30.9 \text{ kN}$$

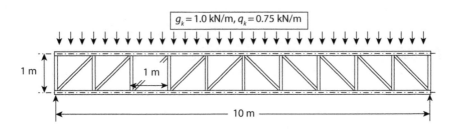

Figure 11.9 Timber flat roof truss.

2. The second moment of area of the top chord is

$$I = \frac{100 \times 100^3}{12} = 8.333 \times 10^6 \, \text{mm}^4$$

It is assumed that lateral buckling of the truss is prevented by the roof. The top chord is restrained against buckling at node points by web members; therefore, $L_{cr} = 1000$ mm and from Equation 11.14, the elastic critical buckling force is

$$N_{cr} = \frac{\pi^2 EI}{L_{cr}^2}$$

$$N_{cr} = \frac{\pi^2 \times 7200 \times 8.333 \times 10^6}{1000^2} \times 10^{-3} = 592 \, \text{kN}$$

And from Equation 11.13, the crushing force is

$$N_{crush} = 7.0 \times 100^2 \times 10^{-3} = 70 \, \text{kN}$$

From Equation 11.12, the compression strength is

$$N_{b, \, Rd} = \left(\frac{1}{70} + \frac{1}{592} \right)^{-1} = 62.6 \, \text{kN}$$

3. The elastic section modulus of the top chord from Equation 11.9 is

$$W_{el} = \frac{100 \times 100^2}{6} = 167 \times 10^3 \, \text{mm}^3$$

From Equation 11.8, the bending strength is

$$M_{el} = 7.0 \times 167 \times 10^3 \times 10^{-6} = 1.17 \, \text{kN.m}$$

The amplification of moment's factor from Equation 11.16 is

$$\alpha_y = \frac{1}{1 - N/N_{cr, \, y}}$$

$$\alpha_y = \frac{1}{1 - 30.9 / 592} = 1.055$$

The UDL induces a moment in the top chord. If it is assumed conservatively that the top chord is pinned at the nodes (because of joints in the timber), then the sagging moment in the top chord between nodes

$$M = \frac{wL^2}{8} = \frac{2.475 \times 1^2}{8} = 0.309 \, \text{kN.m}$$

Equation 11.15 can now be populated

$$\frac{N}{N_{b,\,Rd}} + \frac{\alpha_y M_y}{M_{el,\,y}} + \frac{\alpha_z M_z}{M_{el,\,z}} \leq 1.0$$

$$\frac{30.9}{62.6} + \frac{1.055 \times 0.309}{1.17} + 0 = 0.77$$

Since this is less than 1.0, the top chord of the truss should be sufficiently strong. This solution ignored the moment induced in the top chord caused by the axial force multiplied by the sag induced by the UDL, termed a P–δ moment (see Figure 3.7). In this instance, the top chord sags by 0.5 mm between nodes with web members; therefore, this moment is insignificant.

4. Using the parallel axis theorem, the second moment of area of the truss is

$$I = 2 \times 8.333 \times 10^6 + 2 \times 100^2 \times 500^2 = 5.017 \times 10^9\,\mathrm{mm}^4$$

The unfactored (SLS) load is

$$w = 1.0 + 0.75 = 1.75\ \mathrm{kN/m} = 1.75\ \mathrm{N/mm}$$

And the midspan deflection is approximately

$$\Delta = \frac{5wL^4}{384EI} = \frac{5 \times 1.75 \times 10000^4}{384 \times 7200 \times 5.017 \times 10^9} = 6.3\ \mathrm{mm}$$

This calculation will underestimate deflection, because it ignores shear deflection (caused by the stretching and squashing of the web members). Regardless, the deflection is low and a more accurate analysis would not change the overall conclusion that the deflection is well within reasonable limits.

Problems

Solutions to these problems are provided at https://www.crcpress.com/9781498741217

P.11.1. The cross section of a wood box beam is shown in Figure 11.10. Webs of the beam are fastened to the flanges by screws having an allowable load in shear

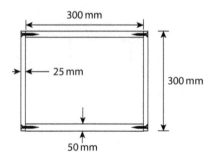

Figure 11.10 Wooden box beam cross section.

of $F = 3000$ N per screw. The beam spans 5 m between simple supports and is subjected to a point load at midspan of 15 kN in addition to a UDL of 8 kN/m, both factored.

a. Determine the second moment of area of the section.
b. Determine the maximum permissible spacing of the screws.
c. Determine the distance from the support that the spacing of the screws can be doubled (i.e. the 50% shear force point).

Ans. (a) 587.5×10^6 mm^4, (b) 68 mm and (c) 1.7 m.

P.11.2. Timber struts prop an excavation, as shown in Figure 11.11. $E = 3000$ N/mm^2, the design stress is limited to 10.5 N/mm^2 and the density of the timber is 400 kg/m^3. Ignore P–δ moments due to the self-weight deflection.

a. Determine the elastic moment capacity of the strut about each axis.
b. Determine the unfactored moment from the strut's self-weight.
c. Determine the buckling strength of the prop in the absence of moments.
d. Determine the FoS if the strut resists an unfactored propping force of 75 kN.
e. What is the maximum sideways force that the strut can resist in addition to the 75 kN propping force?

Ans. (a) 31.5 kN.m, 21 kN.m, (b) 0.80 kN.m, (c) 199.7 kN, (d) 2.48 and (e) 8.3 kN.

P.11.3. A 6 m long beam supports a uniformly distributed load of 4 kN/m dead + 12 kN/m imposed, in addition to an imposed point load of 2 kN at midspan. The section is built up from a solid timber section strengthened with two plates, top and bottom (Figure 11.12). Young's modulus for the timber and steel are 3000 N/mm^2 and 210,000 N/mm^2, respectively, and the crushing strength of the timber is 10 N/mm^2.

a. Determine the ULS moment and shear force.

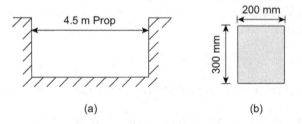

(a)   (b)

Figure 11.11 Excavation propped with timber struts. (a) Cross-section through excavation and (b) cross-section through strut.

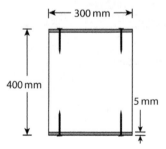

Figure 11.12 Timber beam strengthened by steel plates.

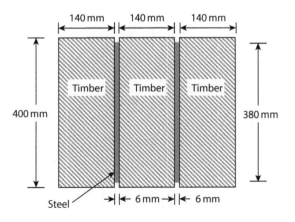

Figure 11.13 Cross section through a flitch beam (connecting bolts not shown).

b. Determine the second moment of area of the transformed section (transform timber into steel).
c. Determine the maximum stresses in the steel and timber.
d. If the plate is connected to the timber section using groups of two screws, each with a shear capacity of 4.5 kN, determine the spacing of the screws along the length of the beam.
e. Determine the distance from the support that the spacing of the screws can be doubled (i.e. the 50% shear force point).
Ans. (a) 110 kN.m, 72 kN, (b) 138×10^6 mm⁴, (c) 159 N/mm² and 2.2 N/mm², (d) 58 mm and (e) 1.54 m.

P.11.4. A flitch beam is made of steel plates and timber joists bolted together to form a single member (see Figure 11.13). The beam spans 8 m between simple supports and supports a UDL of 2.6 kN/m dead and 2.2 kN/m imposed (unfactored). Young's moduli are 8000 N/mm² and 210,000 N/mm² for the timber and steel, respectively.
a. Determine the total stiffness (EI) of the compound member.
b. Determine the midspan deflection under SLS loads.
c. Determine the ULS design moment.
d. Determine the stresses in the timber and steel under ULS loads.
Ans. (a) 2.95×10^{13} N.mm², (b) 8.7 mm, (c) 54.5 kN.m and (d) ±2.96 N/mm² and ±73.7 N/mm².

Index